ABSOLUTE SPACE,
ABSOLUTE TIME,
& ABSOLUTE MOTION

ABSOLUTE SPACE, ABSOLUTE TIME, & ABSOLUTE MOTION

PETER F. ERICKSON

To order additional copies of this book, contact:
Xlibris Corporation
1-888-795-4274
www.Xlibris.com
Orders@Xlibris.com
22751

CONTENTS

TO DR. DEAN TURNER,
WHO REALLY WAS AHEAD OF HIS TIME

PREFACE

The twentieth century is no more. If it had a reigning idea, it was revolution. Everybody has heard about Marxism-Leninism, Freudian psychiatry, the Einstein theory, etc. Eventually, one grew to expect improved household products to bid for the prospective buyers' approval with the claim that they were "revolutionary." In a recent issue of *The Wall Street Journal* appeared the now commonplace headline: "Home Depot's CEO Led a Revolution, But Left Some Behind."[1]

An important characteristic of such a "revolution" is that what is lowermost becomes uppermost and *vice versa*. Then it is supposed that the wheel stops, that it ceases to move on, no longer displacing that which is eminent and putting the vanquished back on top. But that is the defect in the favorite imagery of the so-recently deceased century.

The revolutionary attitude prevailed, not only in politics, psychology, physical theory, and advertising, but also in mathematics, long considered to be the very epitome of certainty. In 1980, a distinguished academic professor in that field had a book published by the Oxford University Press titled : *Mathematics: The Loss of Certainty*. In his preface, he would say: "We know today that mathematics does not possess the qualities that in the past earned for it universal respect and admiration."[2]

At the same time, the mathematical achievements of earlier centuries would take man to the Moon.

Why has it ceased to earn that respect and admiration? Some of the ways which brought it down will be shown in the following pages, but only in passing. The purpose of this book is not primarily to lament the trespasses against reason of the last century, but to travel along the path of restoration and renewal.

On this path, one will not find any attempt to find a profound truth about mathematics in the proposition uttered by the Cretan who says that "all Cretans are liars." This statement does not reveal anything important about the nature of propositions. The contradiction lies only in the content, not the form. If the Cretan had instead said that "all Cretans are truth-tellers," his proposition would have had exactly the same form as the other. It would not have been self-contradictory, but it would still have been false. Its significance is in morality, not mathematics, for it shows that while someone can always tell the truth, no one can lie all the time.

Neither will one find any discussion as to whether or not logic is the foundation of mathematics. People who try to think that way suppose a foundation in analogy with gravity where the lighter rests upon the heavier; either that, or, more commonly, of a hierarchy, such as is found in a system of axioms. But logic is structural, not foundational. It is used in the proof of the secondary, tertiary, and quaternary theorems, as well as in the enunciation of the highest, most comprehensive axioms. It is part and parcel of every element of the lighter as well as the heavier. Where it ceases to be present, the reasoning goes awry. It pervades the fine arts, no less than the hard sciences.

I often disagree with Ayn Rand, but she had well grasped this point when she defined logic as "the art of non-contradictory identification."[3] This is true as far as it goes. Logic is an art, since it is involved in the application or fitting of the end chosen to the appropriate subject matter. But it has scientific aspects, as well; there are important classifications and rules and procedures already known to man. But this does not make it either the root or a branch of mathematics. It is involved in all forms of human endeavor. In general terms, *logic is the art and science of correct reasoning*.

To find this path out of the quagmire of doubt, confusion, and bluff, one must be willing to return from the highly artificial ideas favored in the late 19th century and throughout the 20th century to those founded on sense and natural reason. The typical abstractions of that day were adaptations of Kant, especially his *Critique of Pure Reason*. Generally, the adaptation leaned either in the direction of Plato or of Hume. One or the other was stressed: either the Platonic belief in self-subsisting ideas or the Humean notion that mathematics consists only of relations between ideas and tells us nothing about reality.

Although these modernists reject Plato's notion that material objects and actions are only imperfect exemplifications of the supernal ideas, they retain his notion that sense experience is an unreliable guide. Not all of the moderns were Kantians, however. To give an important example, Georg Cantor said that his transfinite numbers applied to external reality as well as to pure mathematical thought.[4] His philosophical base was Plato, as modified by Leibniz and Spinosa.[5] A position midway between the Platonists and the Kantians was probably that of Gauss who distinguished between an abstract theory of magnitudes and spacial notions which he held to be but exemplifications of the former received through the senses.[6]

The extreme anti-Platonists held with Wittgenstein that one could throw away the ladder after one had reached a certain height and then safely go on. The moderates agreed with the noted physicist, Sir Arthur Eddington, who wrote:

"recognizing that the physical world is entirely abstract and without 'actuality' apart from its linkage to consciousness, we restore consciousness to the fundamental position . . . ."[7]

These men rejected the Aristotelian approach that all ideas ultimately came from nature. This book is neither pro-Aristotelian nor anti-Aristotelian. In the important issue of actual infinity, it sides against Aristotle. With respect to the issue of Plato versus Aristotle, it is close to that famous painting by Raphael, "The School of Athens," in which these men appear as central figures—Plato, with his finger pointing upwards and Aristotle pointing downward toward the earth with the rest on lower levels. But it is not that either.

Altogether, the standpoint of this book is closer to that of the Greek Mathematicians than it is to any of the philosophers just mentioned. Near the end of his life, Gottfried Frege, one of the pioneers of modern set theory, realized that what was needed was a return to a geometrical foundation.[8] This book opens up the way to the new basis through a defense of the infinitesimal and the return of the number line. As stated before, it is not a thematic refutation of either Plato, Leibniz, Hume, Kant, Cantor, Rand, or anyone else, but will simply clear their ideas out of the path when they become obstacles.

This book provides the basis for such a return. That basis is philosophy. The product of decades of search, it began to crystallize while I was writing an earlier work, *The Stance Of Atlas*. The reader does not have to know anything about that other book in order to read this one with full understanding.

Some may think it untoward that someone should talk about a return when American civilization is crashing about us, but here it is.

CHAPTER I

THE PLAN OF THE BOOK

The purpose of this book is to prove the existence of the infinitesimal and reveal some important facts regarding the nature of numbers and the outer infinity, including the existence of the void. Those who choose to follow along will receive some understanding of absolute space, absolute time, and absolute motion.

We shall begin by discussing the nature of the number-line and the basic reasons for its construction. Then we will show the alternatives. The first alternative presented is the notion that not only are numbers finite, but that there is no actual infinity. The inadequacies of that idea will be presented. The second alternative is the notion of infinity held by such as Dedekind and Cantor; it too will be refuted. The actual nature of the infinitesimal and the finite will be discussed; the number line will be reinstated; a type of number very different will be presented; the error contained in the notion of the limit will be identified; and the elements of space, time, and motion will be discussed.

Mathematics could not have existed without men first having discovered two different but related facts, the continuous and the discrete. Aristotle recognized this.[9]

A tiny child notices the changes in his surroundings; that colors and textures vary, that objects move out of one place into another. Such experiences teach him that an existent is one thing and not another and that a single thing can move from one place to the next. Out of this gels a notion of the discrete. That same child notices that a moving object does not fly into myriads of pieces while moving, but somehow remains the same; that what he would later learn to identify as "wall paper" exhibits many repeating patterns of colors. Out of such experiences gels a notion of the continuous.

Various writers have tried to assign priority to one or the other. The important 19th century mathematician, Richard Dedekind, chose the continuous: "The more beautiful it seems to me that man can rise to the creation of the pure, continuous number domain, without any idea of the measurable magnitudes, and in fact by means of a finite system of simple thought steps; and it is first by this auxiliary means that it is possible to him, in my opinion, to turn the idea of continuous space into a distinct one."[10] In the next chapter, we shall encounter a twentieth century school which placed the emphasis on the discrete.

Actually, neither idea can be fully defined independently of the other. To attempt to define "continuity," for instance, one either uses synonyms like "connected," "linked," or negations of the discrete, like "unbroken." The discrete is defined with reference to some change or breach or division or disruption. The truth is that both ideas are defined in terms of each other. It is impossible to define one without being aware of the other on some level. They are correlative ideas. They are discovered together. One is the negation of the other, and yet they cover the whole field of quantity or magnitude.

This is not the only case of such correlatives. An example of correlatives within the realm of the concrete is the following trio: equal, greater, and less. It is impossible to think of equality without the other two; *vice versa*, one cannot start with the thought of the greater or lesser, while having not a smidgeon of thought about that which is the same, or equal.

The ideas of continuity and discretion precede the notion of number. From the side of the discrete, number arises through counting. From the side of the continuous, number arises from measurement.

Children in school usually learn numbers through counting things like little balls and by sorting them out, according to color or size. In the next chapter, we shall discuss the inadequate idea that this is the finite basis of mathematics.

The ancient Greek mathematicians identified numbers with lengths; their emphasis was on continuity. What we characterize as "1" was to them a standard length, which was to be multiplied to find greater numbers, or divided to form fractions. It seems easier to us to use the system we are familiar with, the one that is based on the discrete. We add 1 and 2 to get 3, and we can quickly find that 1/3 of this is unitary 1. Yet, there is an important category of numbers that is completely understood through the Greek system, but only approximately through discrete numbers.

Consider the following drawing:

What is the length of the longest side—the one with the question mark?

According to the Pythagorean formula, the square of the hypotenuse of a right triangle—that is the longer side—is equal to the sum of the squares of the two adjoining sides. Therefore, 1+1 = 2. So the square of the third side is two, its actual length being the square root of two.

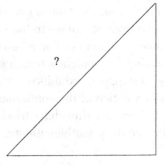

Now, the two adjoining sides of the triangle are each equal to one unit. This unit can be anything we like: feet, inches, meters, miles—whatever. What then is the length of the third side? We have found it to equal √2, but how long is that?

We know that its length exists. We can plainly see that it does. It touches each of the two adjoining sides. Place the length of one of these beside the third side! Note that the third side is a little longer. Exactly how much? Surely, there must be a way by which we can in theory, even if not perfectly in practice, determine how much longer the third side is in terms of feet, inches, meters, miles, etc.

Since √2 x √2 = 2, we suppose that this square root of two must be a distinct magnitude, proportional to the unit 1. It does not matter if it is a long fraction. It is only that it must bear some fractional relationship to the unit one. Let us state it in terms of tenths. Then we can use decimals. Suppose we find it to be 1.4142 times as great. Then we have found that it is equal to one whole unit in the single digit place, to four-tenths, to one-one hundreths, to four-thousandths, to two-ten thousandth.

But the truth of the matter is that √2 cannot be stated in terms of tenths at all. "1.4142" is not the square root of two. It is only an approximation carried out to the ten-thousandth place. The truth is that we cannot find a fraction of any kind, millionths, billionth, etc., that can state the length of this third side of that triangle. This number bears no fractional relationship to the other two sides.[11]

In terms of the seemingly more precise discrete digital system, this length cannot be reached, only approximated. Yet, in the Greek system, there is a length which exactly corresponds to it. Even though we may never be able to locate that precise length with any pointer—a laser would be too wide—still, we know that such a length must exist. But our digital approximation, even if we were to carry it out beyond the ability of science to find it, would actually be somewhat different from √2. The difficulty is not practical, but rooted in the nature of reality itself.

The Greek system of numbers is more cumbersome than the one we use and tend to take for granted. Yet, it is also more exact. Irrational numbers cannot be expressed in whole numbers or fractions. But in the Greek system of enumeration, the position occupied by √2 is a point at a certain distance from the beginning. (In Book X of the *Elements*, Euclid finds irrational straight lines with roots greater than 2.)[12]

Is there a way of accommodating both? There is. Some genius conceived of the number line in which both the continuous and the discrete can be expressed together. (See the illustration on page 16.) Once the unit is defined, it can be multiplied to any length. It can be divided into any rational quantity, any whole number or fraction, and these can be precisely identified. Each would be symbolized

2

1 1/2

1

1/2

0

as with standardized notations, such as 1, 1/3, 1/100, etc. Irrational quantities, although they cannot be determined by discrete numbers, must have their place. Once the unit is set, one knows in advance the point which must stand as the endpoint—say 2—for the interval between 0 and itself, 2. By necessary implication, any number, rational or irrational between these endpoints, must have a position on that line, even though one may never be able to discern which point corresponds to such irrational magnitudes as the square root of two or three. And, of course, the line is capable of indefinite extension.

Since the late 19th century, this solution has been implicitly rejected in favor of others. Replacing this kind of number line is a signature idea of 20th century speculative mathematics. This book challenges that idea.

Henri Lebesgue, the inventor of the integral named after him, endorsed the complete replacement of the number line. Yet, he could not do without it in his elementary definitions. This is implied in his definition of magnitude. "A magnitude G is said to be defined for the bodies belonging to a given family of bodies if, for each of them and for each portion of each of them, a definite positive number has been assigned."[13] Then he defined a body in geometry, as "the meaning of a set or figure."[14] In the same place, he stated that "the length of a segment or of an arc of circle, the area of a polygon or of a domain delimited in a surface, and the volume of a polyhedron or other solid have been defined as positive numbers assigned to geometrical entities and are completely defined by these entities up to a choice of units."[15] Although the theoretician in him yearned to replace the number line, the technician in him leaned upon it.[16]

In the succeeding chapters, it will be shown that the number line itself depends upon the idea of the infinitesimal.

Sometimes, the infinitesimal is defined as the next thing to nothing. But this definition is no more than a characterization. There is nothing next to nothing, except nothing. Any existent, however small, exists. That will do for now. Within any finite extent, there are quantities of infinitesimals beyond measure.

The distinction between the finite and the infinitesimal will be defined in this book. The infinitesimals of time and motion will also be shown to exist.

The question as to whether there is such a thing as an ideal unit length will be answered. Also, the issue of an optimal unit length will be brought up.

Sometimes, the word "infinite" is taken to mean only that it exceeds any preassigned number. This was the meaning intended by the famous mathematician, Karl Friedrich Gauss, when he wrote that "I protest against the use of an infinite quantity as an actual entity; this is never allowed in mathematics. The infinite is only a manner of speaking, in which one properly speaks of limits to which certain ratios can come as near as desired, while others are permitted to increase without bound."[17]

In the chapters that follow, the term "infinite" is taken to mean the very idea against which Gauss made his protest. It will be shown that there is, in fact, more than one kind of actual infinity, as well as the more easily grasped potential infinity.

Since at least the 18th century, the infinitesimal has been shoved aside in favor of the "limit." This decision was made because the limit remains in the realm of the finite. The attentive reader will discover that the idea of the limit is not nearly as rigorous as it is supposed to be, that it leads to a theoretical misunderstanding of some aspects of the derivative and the integral.

Likewise, it will become apparent that the infinitesimal is the key, not only to the understanding of the finite and the infinite, but also to the integrity of scientific induction. The presence or absence of a single, fundamentally imperceptible infinitesimal has made a *practical* difference.

Along the way, the reader will discover that there is a legitimate use for division by zero. The nature of the angle will be made clear, and the identity of the minimum angle will be disclosed; the character of the minimum circle will also be shown. The solution to the famous mystery of $\sqrt{-1}$ will be revealed; it will be found that imaginary numbers do not belong to some occult realm; that there is embedded within them a type of number system not previously known. Set theory will be shown to be inadequate. An unappreciated aspect of the derivative will be brought out which has hitherto unknown connections to physical science. Some of the theoretical difficulties in supposing that non-Euclidean geometry describes some hidden, but actual existence, will be mentioned.

In the course of this book, there will be discussion about two facts of nature which are neither objects, nor actions, nor attributes, nor ideas, but which yet exist. Although these marvels are neither mental nor material, we are unthinkable without them. Can you guess what they are?

And finally, the reader should find restored some of the certainty in mathematics and physical science that was lost in the last two and a quarter centuries of wild speculation.

CHAPTER II

THE REJECTION IN BEHALF OF THE FINITE

In the last chapter, it was stated that the number line is indefinite in length. No statement has been made as to whether it is actually infinite or only potentially so. This will be determined later. But given any finite extent—even the small one from 0 to 2—it must by the nature of the case possess immeasurable points, each of them contributing to some magnitude when measured from the zero position. We spoke of the actual existence of $\sqrt{2}$, rather than a mere digital approximate.

To those who accept only the finite, such a position is a scandal. The argument for the finite is basically this: Take all the entities in the universe, all their actions, all the permutations and combinations of every aspect, separately and together in every way; more precisely, take all the electrons, neutrons, protons, all the units we have not discovered; all their possible interactions, conjunctions, separations, integrations, and divisions: this magnitude, stupendous as it must be, is still something specific. And being specific, they conclude it is finite.

By finite, they mean something that is definite. And by definite, they mean something that is capable of being comprehended in principle by a number established by men. The infinite would have to be something else—either something immeasurable by man or something endless.

The practical reason for their position is evident: If the universe fits into their definition of finitude, it is in principle knowable to man. This does not mean that there cannot be magnitudes too great for man to comprehend directly. To the best established of our knowledge, the greatest velocity is that of light and the nearest stars after the sun are several light years away. But given enough time, all these—especially the problems of man living on earth—could be solved. If there are realms so far away that a man would die even if he were to travel by a spaceship, he would at least have been able to comprehend it with his senses, had he been there. To this way of thinking, the universe is in principle knowable to man.

To the advocate of the finite, the metaphysical idea of infinity, i.e., that it actually exists, is an invalid concept. Such a person might recognize infinity as potential. In the words of one such thinker, "An arithmetical sequence extends into infinity, without implying that infinity actually exists . . . ."[18]

In the 17th century, a case for the finite was made by Thomas Hobbes; in the 18th century, a case was made by Bishop Berkeley. In the century just passed, the most consistent philosophical position the author has ever encountered is the school of Objectivism, founded by Ayn Rand.

Let us turn to an Objectivist essay titled "The Foundation of Mathematics," by Ronald Pisaturo and Glenn D. Marcus.

They begin with the discrete. Imagine, for instance, a couple of children arranging some fish of about the same size into a couple of piles. Suppose one pile is perceptibly bigger than the other. What does this mean? In the case of these fish, it must mean that there are more fish in one pile than in the other. From such experiences, Pisaturo and Marcus reckon, the concept of quantity could be reached; this, they define as "the degree of repetition of like existents in a group."[19]

Is this definition really universal? In the number line, by contrast, quantity is not necessarily repetition. 2 is the double of 1, but $\sqrt{2}$ is not the repetition of anything finite, but only an augmentation of a lesser magnitude that has been set as the standard unit. In the number line, irrational magnitudes are not conceptual exceptions and do not require special handling. Like whole numbers and fractions, they are magnitudes beginning with the start of the line and terminating at a point some distance away. Pisaturo and Marcus are right in thinking that the number line can be used to count discrete objects; but they are wrong in thinking that it can be confined to that use. There is more to quantity than they are willing to suppose.

The authors go on to argue that the components of the two piles of fish can be brought into relation with one another by pairing each fish from one pile to a fish from the other pile on a one to one basis. By doing so, men can see with their own eyes that one pile is unmistakably greater than the other. Furthermore, men can easily understand that, when comparing and contrasting the two groups, it does not matter which fish in group A is faced off against which fish in group B: in any case, it is an act of induction; one does not have to try every single combination of a fish from group A and a fish from group B to understand that so long as the fish preserve their separate identities, the order in which they are counted is irrelevant. The fish are interchangeable for this purpose, and the act of induction is final, not dependent upon some future enumeration. Mathematics, they wisely say, is not all deduction; like all the other sciences, it is a mixture of induction and deduction.

This is the best part of their reasoning. It deserves quotation.

"The basis of induction is finding the *cause* of a particular situation. Once we identify the causal factor(s) in one case, if we find another case where all the causal factors are present, we realize there is no difference in the two cases in our context. The two cases become indistinguishable or interchangeable, i.e., identical in our context. Thus we know the same result must occur . . . . Finding the causal factors is what allows us to pass from the specific to the general, to recognize an entire *open-ended group* of cases as interchangeable. The essence of induction is identifying classes of interchangeable instances by identifying causal factors. Then, if we start with any of the interchangeable starting instances, we must get one of the interchangeable results." [20]

This statement is almost right. Induction is reasoning from the specific to the general. There are other kinds of induction besides causal reasoning.

To return to their account: As long as each fish remains a fish, i.e., does not completely rot away or is not torn apart through human carelessness or seized by some animal, it will retain its membership in the class of the particular pile of fish in which it had been placed. This process of pairing off individual members of groups, they argue, is a process of *counting*. The point they want to make here is that counting preceded the invention of numbers.

A further development on the way to the number system, they add, would be the facing off of the fish with some other kind of existent—perhaps rocks. Doing this would convince the tinkerers that it could work with other things besides fish.

Eventually, someone would do exactly this. This person might take one out of a group of fish and face it off with some stones. With that the induction would spread; and by implication so would the possibility of deduction.

First, the authors had the children face off two different piles of fish. Then, to shorten their exposition somewhat, they conjectured that it was eventually realized that the stones could be used as the standard against which fish and anything else can be counted. It is out of this, they aver, that the concept of number took its rise.

"A number is a group whose quantity matches the quantity of some standard group."[21] Note that their idea of number is dependent upon comparison to some standard originated by man. It has no meaning outside of this. That much is valid, but note also the implicit bias toward the discrete.

Summarizing the way numbers can be generated from considerations of the discrete, they write: "Observe that counting precedes numbers conceptually. Men needed some primitive form of counting to arrive at specific numbers, let alone the concept 'number.'"[22] Pisaturo and Marcus hold that there must be a face-off comparison of groups before there can be numbers. They believe that numbers owed their origin to the use of the one-to-one correspondence, that numbers arose from the intelligent use of this principle.

This may or may not have been the basis for the invention of numbers. But it is not the basis for the discovery of abstract quantity itself. Before men could conceive of the facing off process, they must first have had some notion of quantity; otherwise, it would be random. Before the facing off could begin, they would first have to put the fish in piles, which means they must begin by concentrating a quantity of something—in the case of their example, a quantity exhibiting the qualities of being fish.

Pisaturo and Marcus would not disagree with that statement. They define quantity as the "degree of repetition of like existents in a group."[23] Their definition makes it plain that they are with the school which holds that mathematics arose out of contemplation of the discrete.

But one can also arrive at the notion of quantities by starting with considerations based upon the continuous. Suppose someone picks up a stalk of grass and bends it into two equal parts and then into four equal parts. Right there, the notion of the many out of the one comes about. It does not take much to see that further divisions are possible. The idea that these divisions can extend beyond the ability of the reed to be bent without breaking could come next. There is no requirement for any face-off. With more contemplation, both the idea of multiple units and also the idea of the fraction can appear. And if this person picks up another stalk and bends it into five equal parts, the notion of the even and the odd is there for contemplation.

Pisaturo and Marcus require the group as a conceptual background against which to identify the separate units. The green stalk would serve just as well for the abstraction of the idea of individual numbers. Repetition of distinctly different objects of a same kind would not be required. It would be enough that they issued from the same source. The truth is that mathematics is founded upon two central ideas: the continuous and the discrete. It may have begun from either.

Now, let us return to what Pisaturo and Marcus say about counting. In this activity, they argue, each member of the group is counted as if they all were the same, even when they differ, perceptibly, with respect to such attributes as weight and size. In the example of fishes, although one can see that they are not identical in all respects, they are, each of them, units.

But mere counting is not measurement. To remedy this lack, Pisaturo and Marcus ask that we imagine an early explorer who would have needed to carry with him enough venison to last for ten days in the wilderness. Since he would have to carry it and more on his back, economizing the burden would be important He might take out a large chunk that he knew would work for a certain part of the journey. He knew this by feeling its weight. Then using his hands as a relative scale, he would take out pieces of the remaining meat which when added together with the one he was using for a standard would weigh enough for the ten day trip. Not every piece would do. Some would be too heavy, some too light. He would be using the first chunk as the standard, and it would be the unit by means of which he would *measure* the weight he would take with him.[24] "Measurement," as Ayn Rand wrote, is "a standard that serves as a unit."[25]

This, they argue, shows how one can get from the counting of the fish, which is purely qualitative (in that every fish counts as the same) to the quantitative, where scientific measurement is possible. The crude example of the chunk of venison, isolated by means of the attribute of weight, is, of course, rendered by them with greater refinement than has been stated here.

Note that by this argument, they have subtly passed from the discrete to the continuous. The standard weight would be discrete, but the size of any chunk cut out of the whole would be dictated by the idea that the part of the animal's carcass which the explorer was chopping up was continuous with respect to the survival food value for the trip. In the example given, any piece of a certain weight—the standard serving as a unit—would do. Differences in color or texture, etc., would be ignored. In that context, the pieces are interchangeable.

Then they announce their grand conclusion: "The premise of the interchangeable unit is what we may call the axiomatic condition of mathematics; when the condition is true, mathematics applies."[26] Yet, we have found that there is an element of continuity in the very example which they have given. The part of the carcass from which the explorer cuts must have the potentiality of having chunks of venison of the same serviceability drawn from it. Originally, the individual chunks were connected to each other. And after they had been separated from each other and laid in the knapsack, there was no breech in quality as far as the explorer's requirements were concerned.

As was stated before, the discrete and the continuous are correlative concepts which require each other for mutual intelligibility. To return to the counter example of the man bending up the stalk: if the ancient investigator had wished, once the stalk was bent into various parts, then the parts could have been separated and afterwards treated as discrete objects. The notion of the continuous as a basis for number creation might have preceded the use of the discrete in this regard, rather than the other way around.

Pisaturo and Marcus continue: "Let us now translate the condition of the interchangeable unit into a axiom: $1 = 1$. Observe that the concepts of quantity and unit are contained in this axiom. The concept of counting and number are also contained in it: the symbol used for a unit is the number '1', denoting a *counted* unit.

"This axiom of mathematics is *not* a restatement of the axiom of identity, 'A is A.' 'A is A' means a thing is what it is. The statement '$1 = 1$' is not merely an attempt to restate 'A is A' in the form a thing equals itself. Rather '$1 = 1$' means that each thing in the group equals each *other* thing in the group; each 'one' equals itself. This axiom is not, most fundamentally, a statement about numbers; it is, most fundamentally a statement about *units.* It states that every group of one unit within a larger group of units equals every other group of one unit within that group of units."[27]

But is this pithy formulation correct? Consider the counter-example of the folded stalk. It is certainly a unity. It is not only equal to itself, *qua* unbroken stalk, but it is equal to the sum of its parts. It is neither a unit that serves for a standard nor a standard that serves as a unit in the sense outlined by Pisaturo and Marcus, but it is a unit in the sense that it is one out of which many are derived. It also stands as a basis for calculation; its parts can be counted; they can be enumerated. The process used is division, a refinement of subtraction, rather than the addition and multiplication these writers have in mind.

Reflect upon only the first division of the stalk into two pieces. There, subtraction would be the process of finding the difference between the whole stalk and one of the folded parts, which would, of course, be the length of the other one. Division would be the process of finding what part one of the folded parts is of the whole stalk, which in that instance would be *1/2*.

Pisaturo and Marcus suppose that addition is more fundamental than subtraction. But this is really optional. A person could start with the stalk and after making the subdivisions through division, reason backwards and come up with addition. The two processes were reversible—at least until the discovery of zero, a much later refinement.

Note further, that in the instance of the folded stalk, the units are not really interchangeable, although they are of the same size. In order to exchange a unit from farther down the stalk with the one at the tip of the plant, the stalk would have to be torn in pieces, which would destroy its unity. After that, they could be added together in any way, as is supposed to be indispensable by these Objectivist authors. This shows how much more versatile it is to begin with the stalk.

Sir William Hamilton once argued that from the standpoint of geometry, subtraction is more fundamental than addition. His words are worth quoting here:

> "To illustrate . . . let SPACE be now regarded as the *field* of the progression which is to be studied and POINTS as the *states* of that progression. You will then see . . . that I am led to regard the word 'Minus,' or the mark -, in geometry, as the sign or characteristic of the analysis of one geometrical position (in space) as compared with another (such) position. The *comparison of one mathematical point with another*, with a view to the determination of what may be called their *ordinal position* in space, is in effect the investigation of the GEOMETRICAL DIFFERENCE of the two points compared, in that *sole* respect, namely *position*, in which two mathematical points *can differ* from each other. And even for this reason alone, although I think that other reasons will offer themselves to your own minds, when you shall be more familiar with this whole aspect of the matter,

you might already grant it to be *not unnatural* to regard, as it has
been stated that I *do* regard, this study or investigation of the relative
position of two points in space, as being that *primary geometrical
operation* which is *analogous to algebraic subtraction*, and which I
propose accordingly to denote by the usual mark (-) of the well-
known operation last mentioned."[28]

Interestingly enough, Pisaturo and Marcus do consider ordinal numbers.
They argue that the primitive ability to count does lead us to the notion of such
numbers. In ranking piles of fish, one can say that one pile is greater than
another, but, they add, such a comparison is insufficient for scientific purposes.
Yet, Sir William Hamilton has shown that it can be used to great effect in certain
elementary geometrical studies when subtraction is used first.

Actually, it would not matter if everyone started with addition, rather than
subtraction. What matters is that it can be done the other way, for the two processes
are correlative.

These authors add: "That the axiom '1=1' is implicit in all mathematics can
be seen by trying to devise a mathematical statement that does not assume it."[29]

This is true as far as it goes, but it does not decide the issue as to whether
the idea of numbers came out of the discrete or the continuous. They wrote
that "'1 = 1' means that each thing in the group equals each *other* thing in the
group."[30] But consider the situation when the group is not the central idea:
There, the stalk of grass does not have to be divided into equal parts. It can be
bent into various lengths. If it is bent into two parts and one part is greater
than the other, they do not have to be treated as if they were the same. There
are still so many parts. The person doing the dividing would not suffer confusion,
because he can could see with his own eyes, or feel with his hands (if he were
blind), that each of these differing divisions is part of a greater whole. Referring
to a particular division—say the first one formed—where the two parts were
not of the same length, the original unity can be reassembled out of the two
unequal divisions.

New units can be created out of this assemblage. Suppose that the smaller
of the two parts is in turn bent into three unequal sub-lengths: together, these
three equal the entirety of this second 1. This 1 is also equal to the sum of its
parts, and of course, itself. Here, 1=1 does not mean that each sub-unit is the
same as the other two. Here, it means simply that the subunits added together
equal that out of which they came with respect to length.

Pisaturo and Marcus stress that their 1=1 does not mean the same as A =A
in logic. Yet, the one as unity deriving from the contemplation of the divided
stalk means that a thing is equal to itself. Their notion is founded entirely upon
considering the discrete. (Later on in the exposition, these two slightly different
concepts will be identified as the 1 of identity and the 1 of unity.)

What do Pisaturo and Marcus say about geometry? "The concepts of quantity, counting, and number (and, therefore, the axiom of mathematics, '1 =1') are not implicit in many geometric statements."[31] From this, they conclude that geometry is not a branch of mathematics.

These are the examples they give as to where geometry does not fit into their concept of mathematics. "'Straight paths that do not share a plane will not cross.' 'A circle can fit inside another circle of greater diameter.' Moreover, geometric statements rely on *other* basic concepts, such as 'physical extension,' 'direction,' and 'shape.'"

Contrary to Pisaturo and Marcus, the number line uses negative numbers, which involves direction; concentric circles offer it no problem; neither does a line in a second plane. Their over-emphasis on the discrete causes the field of mathematics to shrink from its long recognized boundaries.

They wish to confine mathematics to discrete quantity and exclude space from it. "Geometry deals with only one kind of uniform unit: units of physical extension, i.e., length. On the other, mathematics deals with general principles about units that are uniform in *any* attribute—weight, length, time, function, etc."[32]

By excluding geometry, they automatically cut themselves off from the number line as an indispensable basis for the articulation of the science of numbers. Their strategic purpose, as will be shown in the next chapter, is to obviate such ideas as the infinite and the infinitesimal.

CHAPTER III

IN DEFENSE OF THE IMMEASURABLE

In common speech, two kinds of infinity can be found, actual and potential. What is now called "potential infinity" was defined by Aristotle thus: "A quantity is infinite if it is such that we can always take a part outside what has been already taken."[33] This is a type which is thought never to reach completion. The best known examples are asymptotes where the curve approaches a line which it may never reach; that, and geometric series like: $1/2 + 1/4 + 1/8 + 1/16 + 1/32$, etc. which cannot on those terms ever add up to 1, but can get as close as one wishes. A more recent example is the astonishing "fractals," images within images, which are inherently unfinishable, but which exhibit recognizable over-all forms.

Actual infinity refers to that which is either immeasurable, endless, or both. The reader will come to a thematic discussion of this in Chapters VIII and IX. In the present chapter, it is one of the two kinds of actual infinity, the *immeasurable*, that will be defended.

Aristotle did not believe that actual infinity exists. Modern finitists like Ayn Rand and her Objectivists agree. Infinity is potential only. Reality is finite. "An arithmetic sequence," she writes, "extends into infinity, without implying that infinity actually exists; such extension means only that whatever number of units does exist, it is to be included in the same sequence."[34]

In the case of the sequence of whole numbers, she is technically correct. It is finite. Any single number, however great, must be finite. Infinity cannot be a number; for if it were, then a bare infinity minus one would be N, a finite number. As Euclid once proved, N + 1 cannot be infinite. It can only be finite. But it does not follow from this that there is no actual infinity.

Let us begin by examining the idea that there can be a potential infinity without there actually being any infinity, *per se*. Can there be a process with no end and yet have the idea of the "no end" be a spurious reification?

Consider the common idea that the curve can never meet the asymptote. This "never" is not like that of 2 + 2 = 5 in which the proposed sum is simply wrong, once and for all time. Neither is it like the statement that a certain bird which was killed by a cat will never eat another meal. A potential infinity is thought to continue, but not conclude. At the extreme, this implies an infinite amount of time. To contradict this, one would have to show that there can be a process which involves no time. But that would be a contradiction in terms; a simultaneous coincidence could not be processional.

Yet, this is what the advocates of the finite would have us believe. They would have us accept the incomprehensible absurdity of a timeless process in order that they not entertain the idea that something might exist which is beyond their ability to fully comprehend.

Consider in a general way the supposition that potential infinity exists but actual infinity is a contradiction. Why then even make reference to it? The "extension" Rand mentions is supposed to go on and on, i.e., without end; the notion of an actual infinity is referred to, although not acknowledged. Can this be reason? How can a person intellectually learn a non-existence and still be correct? Why then indirectly refer to something which one does not believe in? If the notion of an actual infinity is absurd, can one have a concept in a reasonable operation that is itself a contradiction? Can one have a standard that cannot be? What a scandal that is to a philosophy that purports to be based upon reality. Here, we have non-reality serving as a guide for the student of reality. Elsewhere, when an idea has been shown to be fallacious, it is no longer used in the process of discovery. Workers in thermodynamics do not use Phlogiston as a tool for the study of heat, any more. Astronomers do not use the avoidance of the Ptolemaic epicycle as a means for calculating the orbit of the earth. Physicists do not use Aristotle's dynamics as a negative guide in constructing vector equations. If there really were no such thing as the infinite, why is it required, even as a potential? A potential that is self-contradictory is no potential.

Obviously, there is something wrong with the reasoning of those who follow Aristotle (or Rand, or whomever) in this regard. As we shall come to understand, it is the fact that actual infinity does indeed exist.

Infinity is a quantity so great that it is qualitatively beyond the finite. That is why it is called, **"Infinity."** The reason why it can only be expressed in a negative form—contrary to the standard that a proper definition be positive[35]—is that our actual experience is only of finitude; every sensation we receive is specific and is grouped into definite perceptions. Yet, as was shown above, we can form such notions as these, even though we cannot understand their full philosophical ground; more to the point, not only are we able to form them, but in certain situations we are required by the very nature of the conjugations of the finite to do so.

To this, the denier of the actual infinite might answer: Let us return to the number line; that line must always be finite in length. No matter how many whole multiples of the fundamental unit 1 are made, for the reasons given above, that line can only be as long as it is. Lines are extensions in space; and in order to be extended, they must be increased by an action on the part of some extender. The natural numbers and the integers are always finite. No matter how great some number N may be, $N + 1 \neq \infty$. Simply because the number line may be extended without limit means that its development is always finite, never actually infinite. The same is the case for the asymptote. The curve can never meet it. It is a potential infinity.

The advocate of *infinity as the immeasurable* might answer: Instead of looking at the length of the longest lines, let us consider the interval between 0 and 1.This is a finite extent of the number line. Recall the geometric series $1/2 + 1/4 + 1/8 + 1/16 + 1/32$, etc. Each of these corresponds to a position on the number line. Modernist mathematicians agree that the limit of this series is 1. In order for a position to exist on the number line, it is not necessary that it be identified by some human being. The point corresponding to 1/128 would exist, even if one did not bother to calculate that it is the second term in the series after the 1/32 published above. Within this finite length set by the two end points, 0 and 1, there already exist immeasurable positions. When we divide a line, the sections in which it is to be divided must already exist. Each section is composed of innumerable positions. These positions on the line are not potential. They exist. If they did not exist, the line would not be continuous. *But the line is continuous.* If it were not, then there would be some positions that would have been on some other line, but are not on this one. Then *that line* would be continuous. The same reason which makes it possible to know that the length $\sqrt{2}$ exists in a certain region within an interval of 2 units (even though men may not be able to physically locate that point) enables us to know that the as yet undiscovered points corresponding to the as yet uncalculated terms of the series must exist. Just as the Pythagorean formula guarantees the existence of that famous square root, so the formula for the geometric ratio guarantees the existence of any term in the series.

Against this argument, the finitist could frame the following answer: that while the existence of innumerable points within an interval is a fact, there is still an element of potentiality. The intervals corresponding to a point and its distance from the origin must first be calculated in order to have more than a potential existence. Until and unless a number corresponding to some specific point has been established, its existence is no more than a possibility. To have a full existence, it must be calculated. This, of course, also holds for numbers greater than those not yet known: A line which is to be extended must be made to reach to where it was not before; and when this has been done, the process of identifying the new points of numbers has to begin anew.

To such a finitist, the following answer can be made: the term "potential" is being used ambiguously. The use of that term in potential infinity is something men cannot finish. Now, it is being applied to positions which have yet to be named. Once the number line has been established and the unit is defined in terms of its length, the points, whether named or unnamed, necessarily exist within that part of the line which has already been extended. Take a line that is just five units in length. The points within it have relationships. The point corresponding to a rational number like ½ is easy to discover. The point corresponding to π exists within the fourth multiple of the fundamental unit. Although we know that we cannot find it, we know that it must exist, and this knowledge makes it legitimate for us to approximate it. If we didn't think that it existed, there would be no reason to try to get near it through calculation.

Yet, the finitist is not wrong in all respects. The process of calculation within the interval 0 to 1 is beyond human calculability. One cannot even identify all the rational points. Prime numbers are all finite, but innumerable. The task of finding new ones gets more difficult all the time. On June 1, 1999, it was reported that a prime with a million or more digits had been discovered.[36] Suppose some prime number way beyond any number which our number line has so far reached—something way beyond the googled google. Now, according to standard mathematics, every prime must have a reciprocal; this being premised, so must the point corresponding to this fraction exist in the interval between 0 and 1. Therefore, on the finitist's own premise, we cannot even say that every number within the interval of a unit is identifiable.

Let us grant that numbers are an artifice; that they are irretrievably finite. But the points on a line which men mark as 1, 1.1, √3, etc, are not numbers all by themselves. The number is the length from the beginning of the number line to that point. "1" is only the identification of a unit's terminal point, the unit itself being the distance *between* the terminal point and the initial point on the line. The length of the interval is relative to the length of the unit which defines the calibrations of the rational and irrational numbers.

Pick up a ruler. Run your fingers down it from beginning to end. Without identifying anything other than the beginning and the end, your fingers have passed by points beyond number. Under a microscope, the ruler may show great irregularities. An ideal line is not needed here; the points of a crooked line tracing the rough edges of that physical ruler would also be beyond man's ability to calculate. When one thinks of an ideal straight line in space cutting across those irregularities, the point, so to speak, is even more obvious.

Quantity and number are not the same. Numbers can only be finite, but between the endpoints of any unit, a quantity beyond number must exist. Unlike material things, they can not be discerned, even by an electron microscope. No counter can find them, such as is used to detect atoms. They cannot be found by the smallest finite measure.

These are the infinitesimals. They are not zeros. While an immeasurable quantity of them would be required to produce any finitude, however insignificant, an infinity of zeros would still be zero. The best example of the infinitesimal is the point in space. This was defined by Euclid, who said that a point has no parts.[37] One can see why. If it had any parts, then those parts would be the points, and so on.

Sometimes, this is subjected to ridicule. For instance, the mathematician, Friedrich Waismann, once objected that "a pain has no parts; now is a pain a point?"[38] The quick answer is that pain does indeed have levels of intensity. But Waismann was clearly talking about some threshold, some essential minimum of unpleasant feeling. One could speak of this as a "point of pain" in a colloquial sense, but not in the way that one would speak of a point in space or of time. The latter are below the finite; they would not be perceivable even under strong magnification.

These points have locations but no finite length. If they were finite, they would have a length. But if that were the case, then the geometric series referred to above could not be represented on the line. To uphold their doctrine, believers in universal finitude would have to hold that the points were finite in size. But by doing so, they would be undone by the fallacies which confounded Hobbes and Bishop Berkeley.[39]

John Stuart Mill once argued that all that takes place in a geometrical argument is that we abstract the length of a line apart from its thickness and a point from its size or area.

Mill's contention is that we can consider the point as if it had no area, although it must have some area. But if that were the case, why is it that perfectly reasonable conclusions can be drawn from it? Ordinarily, when we ignore some part of reality, we soon have to reinsert the part denied into our thought in order to prevent ridiculous conclusions. Yet, when we try to give lines breadth and points, area, conceptual errors result. If something is really an absurdity, it cannot even be approximated. Who would attempt to approach a square circle with a material representation? It is because everyone knows that it is impossible in space that it is not attempted—or at least, no one advertises the fact that he or she had tried.

Put differently: If one can attend to an attribute like finite length by itself without getting into contradiction, and the attempt to consider both finite length and finite breadth together validates some important geometrical theorems, we are entitled to think that finite length in its pure form can exist in space. And if one dimensional figures are fully intelligible, then so is the point which is required by lines when they meet.

Against this, it might be argued that perhaps, there are only potential infinitesimals, i.e., that one can only divide forever without involving points. This seems to have been what the famous French mathematician, Augustin-Louise Cauchy, was driving at when he said that the infinitesimal was a variable with a limit of 0.[40]

Let us examine this argument: Here it is said, that one can only have intervals, each one smaller than the one before, but intervals, nonetheless. Yet, if we examine the ends of an interval, we realize that each end is a point. Since there cannot be an interval, i.e., a finite extent, without at least two points, the interval requires these points. They, at least, must be solid, for if they could also be divided within themselves endlessly, then there would be no actual size to any interval, only a kind of smear. And if that were the case with length, then there could be nothing definitive to mathematical measurement, even on the theoretical level. But if the endpoints are stable, so are the other points within the interval, for intervals can begin anywhere.

If there could be intervals without end-points, there could be no such thing as an edge. There could be no point at which it ended. This would not only be the case with the material edges, like razors, which show irregularities under a powerful microscope, but even those found in theory. There could not be such a thing as a unit length, for no length could ever be the same as any other length.

The irreplaceable character of the point can be seen in the case of a tangent lying straight on a circle. There is no room for any other line between tangent and curve. This follows from the definition. The diameter of any second line which someone would try to insert in there would be a point. And since there can only be one point at that location, it is not nothing; between zeros, one can always place another zero. Therefore, its non-zero nature is evident. (It is not negligible either, for an important part of the infinitesimal calculus is founded upon another type of non-trigonometric tangent.)

Among other things, the upholders of the finite point would have to deny the existence of irrational numbers. Suppose there were 10^{100} finite points in the interval between 0 and 1, then the smallest "number" would be $1/10^{100}$. All numbers would be rational. Furthermore, various number lines would contain different finite numbers of possible fractions, depending upon the length of the unit interval.

And by denying that $\sqrt{2}$ was an irrational number, they would also deny both algebra and geometry, since the length shown to exist in that famous use of the Pythagorean formula is demonstrated to be not a fraction by elementary algebra, which in turn is founded solidly upon arithmetic.

Irrational numbers cannot be stated in whole numbers or fractions because their magnitude is inexpressible in terms of numbers based solely upon the unit. They touch a level beneath that of the finite.

An infinitesimal has location only; it is not a region to be divided into parts. An infinitesimal is not a number either. It cannot be approximated by a tiny number, defined by $\pm \in$, however small the latter is in comparison with some other number. The actual infinitesimal, therefore, is not the same as that spoken by mathematicians who define it as a "variable which is ultimately to approach the limit zero . . . ." ; it is a constant, but not a finite numerical one.

Let the symbol for the infinitesimal be: "●".This symbol is formed by taking a period and making it as large as a letter.

Let us not conclude from this that the infinitesimal is "round"; these locations have no shape. They are but elements of continuity in space. The instants of time are also infinitesimals, but for now, the spacial point will be used for purposes of illustration. The finite, it should be understood, stands between the infinitesimal and the infinite.

Some mathematicians have attempted to get away from the nature of the point. Thus we find Stephen Barr in his *Experiments in Topology*: "But if a point, as the geometry books used to say, has position but no magnitude—no size in any direction—to say that another point, also having no size, was *next* to it would imply that there was *no distance between them*. This would mean that their positions were the same, and since the only thing that distinguishes one from the other is its position, it would mean that they were the same point: we have consequently failed to put down a *next point*. This is hairsplitting with a vengeance, and important in topology."[41]

The answer is that a point has only location, not area. It takes up that position and none other. Between any two neighboring points, none can be inserted. They are neighbors. But, although they are next to each other, they are not identical.

There is no center to a point, for if there were, the point would have a periphery and would therefore have a magnitude across it. There cannot be distance between one point and its neighbor; for, if there were, it would also consist of points. But, there are none. It is simply a difference—the *difference of a position*.

Being a position, a point does take up space. But it occupies no length; neither do two, nor does any finite number of such points. The infinitesimals within a finite extension are innumerable.

To understand this, one must distinguish between a finite distance or interval and a finite number of neighboring infinitesimals. Take an interval. On each end is a point; between them is a finite distance. Next take a sequence of infinitesimals below the level of the finite. Let us label them 1, 2, 3, 4, etc. The sequence can even be fractionated. A sequence of twenty infinitesimals can be divided into 20 singles, 10 pairs, 4 fifths, or 5 fourths. These are rational fractions. But it can contain no irrational magnitudes. For that would split an infinitesimal into parts, which is impossible.

Why would it? An irrational number is not an aliquot division of a finite unit; there is, for example, no rational number to which the square root of two corresponds; no finite fraction multiplied by itself can yield two. It can never be stated accurately in terms of decimals. The finite basis upon which the unit can be rationally divided does not and cannot measure its end point. That point, although inside the unit, is outside of its purview, of its range of calculation. Alternately stated: the irrational magnitudes cannot be found by using the unit as a standard.

Irrational numbers do not rupture the unit; without them, the unit would not be continuous. But being incommensurable, they show the reality of the realm of infinitesimals which lie beneath the sheath of the finite. In a manner of speaking, an irrational number forces the person doing the calculations to look beneath the level of the finite and see that there is something else, the infinitesimal.

In short, the finite level must contain irrational magnitudes; the level of the infinitesimal cannot. The presence of the irrational is indispensable to any length. That is why a sequence of infinitesimals cannot have a finite length. They are two different levels of reality. This distinction and its importance will become clearer as the exposition proceeds.

The spacial infinitesimal has location, but no parts—Euclid's point, in other words. This infinitesimal is beyond number. Take any interval. This interval is finite. It has a number. Given signed numbers, once a 0 point has been picked, then an interval can be measured. The infinitesimal at the end of the interval opposite to zero has a number also, for instance, $+\sqrt{2}$. But that number, that evaluation, exists only because of its place in that extent is determined by the placement of the zero. Indeed, that same point could itself be chosen as position 0, with the point presently marked by 0 being labeled as $-\sqrt{2}$. In short, the necessity lies in that which must exist at the moment that the 0 point or beginning has been asserted.

The transition from the infinitesimal to the finite is *qualitative*. Bertrand Russell incorrectly supposed the following to be a property of the concept distance. "If A_0, A_n be any two points, there exists n-1 distinct points (whatever integer n may be) on the straight line $A_0 A_n$, such that the distances of each from the next, of A_0 from the first, and of A_n from the last, are all equal."[42] (Italics deleted.) Quite simply, Lord Russell failed to distinguish between the finite distances within an interval and the differences in position between infinitesimals within an interval.

Pisaturo and Marcus say they want mathematical units that are interchangeable in any attribute. They mention "weight, length, time, function, etc."[43] The points in the number line can be used for all of these, even time. The unit which they have come up with is not independent of any other characteristic. It is designed to preclude infinity. They want the unit to be uniform. The points on the number line can handle that requirement quite well. They say that "a large number has an actual referent only if there is a group with that many units in it."[44] The number line has its infinitesimals and its intervals.

Leaving aside the issue of infinity: is it true that whole numbers are best thought of when viewed as members of a group? Consider the numbers 3, 5, 7, 10. These numbers are not members of a group in the same sense that so many wooden balls belong to a croquet set. Three is included in ten; and so are five and seven; three and five are both smaller than seven, and three fits into five. The smaller numbers in the group are actually components of the larger. They are different, but they are best viewed as segments of each other, save for the greatest one. That is the difficulty of thinking of them as parts of a group. One might think of a "set" of four separate balls with diameters 3", 5", 7", 10" as members of a group because of their physical separateness, but not numbers in which the larger cannot exist without the smaller. The number line is superior in this respect also, for it shows not only the distinctness of the separate elements, but also how the larger comprehends the smaller.

In this chapter, the existence of actual infinity has been established with respect to the immeasurable. The realm of finite measurement has been defined as that in which rational quantities can be stated exactly in discrete numbers, but irrational quantities cannot. There is a discrepancy between an irrational number and what can be rationally determined by the unit.

Since irrational quantities must occupy objective lengths in the number line, the finite simply cannot be equated with the definite, as in the previous chapter. This refutes the thesis that only the finite exists.

Said otherwise: As a unit may be defined as a finite interval in terms of which its parts are to be identified, this poses a problem. Not all of its parts can be fully identified in its terms.

In the next two chapters, the modern view which accepts actual infinity, but at the price of endangering reason, will be discussed. After that, the positive case for the infinitesimal in many of its major ramifications will resume.

CHAPTER IV

ON CONTINUITY

The dominant schools in 20th century mathematics rejected the geometrical number line. Although they accepted actual infinity, their rationale was quite different from that made possible by the number line. That will be the subject of the next chapter. In this chapter, the subject is their handling of the concept of continuity.

Up until the 19th century, only the wildest skeptic doubted that the number line is continuous. But certain mathematicians did not want to depend upon this idea. This probably reflected the desire of those who had accepted the radical subjectivity of Kant, but wanted to go beyond him. This attitude can be found in the work of the early 19th century German physicist and mathematician Karl Gauss, who wrote: "But the true content of the whole argument belongs to a higher domain of the abstract theory of magnitudes, independent of the spacial, the object of which is the combinations among magnitudes linked by continuity . . . ."[45] Another German mathematician, Richard Dedekind, wanted a mathematics that was purified of any dependence upon sensual perception. This new mathematics was not to be based on geometry, but instead built out of the materials of the imagination.[46] He was convinced that mathematical concepts are created by the human mind.[47] In a private letter, he wrote, "we are of divine lineage and there is no doubt that we possess creative power, not only in material things (railways, telegraphs) but quite specially in mental things."[48] Like Gauss, Dedekind was a Kantian. Dedekind's some-time collaborator, Georg Cantor, was more of a Platonist, but he too believed that mathematics consists of rigorous deduction from premises.

It would be a mistake, however, to think that they were only trying to be novel or *au courant*. They believed that the traditional understanding of real numbers based upon the notions of geometrical magnitude was deficient, that it did not account for complex or even negative numbers—that, in the words of historian of mathematics, Domínquez José Ferreriós, "the continuity of R was neither justified nor explicitly required."[49]

Central to a system based upon definitions alone would have to be that of continuity. The type of definition which they accepted became the standard in the 20th century.

It was not supplied by either of these mathematicians. The modern standard notion of continuity is that given first by Bernhard Bolzano, namely that f(x) is continuous within an interval if for any x in that interval, the difference f(x +ω) - f (x) can be made as tiny as one can wish by making ω sufficiently small.[50]

The operative assumption is that there is no limit as to how small a subsequent interval within such an interval can be, i.e., that the process of subdivision is endless. If there were a point beyond which it could not be carried on, the definition would be false.

Contrary to Bolzano, continuity cannot be properly defined in that way. Continuity is the state of not being broken, interrupted, or changed in some irregular way. It is defined in terms of its opposite, the discrete; and the latter is defined in terms of it. Both are disclosed from perceiving the world. Yet, although both are abstracted from positive experiences, they can only be defined by contrast. There is no scandal in this. The one lacks what the other possesses.

Suppose, *impossible*, that some person had never heard of continuity. How would he notice it? Bolzano's definition would be of no use. To visualize it, one would already have to imagine a sequence, a notion unimaginable without having thought of one existent following another without a gap, i.e., without having already formed the concept of continuity.

To see this more clearly, let us take the modern definition of continuity in its most abstract way, the formulation of the German mathematician Karl Weierstrass, which does not speak of the interval becoming as small as one can wish, or other subjective expressions to that effect. In the words of renowned mathematician and historian, Morris Kline, "Weierstrass attacked the phrase, 'a variable approaches a limit' which unfortunately suggests time and motion. He interprets a variable simply as a letter standing for any one of a set of values which the letter may be given. Thus motion is eliminated. A continuous variable is one such that if x_0 is any value of the set of values of the variable and δ any positive number, there are other values of the variable in the interval $(x_0 - \delta, x_0 + \delta)$."[51]

Again, if one had no idea of continuity in the first place and had to start with the two number comparison test, then no matter how many times one applied it, one could not be sure that it was not something else. It would be rather like the problem with induction from simple enumeration, the type passed over by Pisaturo and Marcus: just because the only swans one ever encountered were white does not mean that a black swan could not exist. It is because one already understands what continuity is that one might be tempted to reckon this as a proper definition.

Continual divisibility may be a characteristic of someone's idea of continuity, but it is not a proof of the same. A simple discrepancy within an huge assemblage might never be found. Imagine a simple thread of yarn in an "infinite" haystack; since the strands of straw were themselves beyond number, an extensive search might never obtain the piece of yarn. Or, if this example is too surrealistic, imagine the modern idea of the real numbers between 0 to 1 in which every successive number was digitally greater than its predecessor but less than its successor according to all the rules, but that somewhere in the midst of which a discordant "2" had been set. Since the sequence is supposedly endless anyway, one might never encounter the exception using Weierstrass's formulation.

It is only because one already understands what continuity is that one can appreciate what Weierstrass and his followers are talking about.

Even if one accepts the existence of the property of always being able to find smaller intervals within a given interval less than a unit, it is not as universal as the idea of continuity itself. A finite neckless made of intersecting rings of the same size would have a continuous pattern—would it not? And there would have to be a mathematical description for it. The same thing could be said of a fence in which every third post was blue and the rest were white; not only did the posts of that color follow each other without a single breach of that order, even the interruptions were regular and so continuous in a certain way.

According to the original understanding of continuity, a continuous thing is one without breach. In these terms, the sequence of whole numbers, 1,2,3,4 . . . is continuous, since it cannot be interrupted by anything else. As long as they remain whole numbers, nothing can be inserted between them. If we thought of each of these magnitudes as something like a row of dominoes, then there is no way that any different number could be inserted between any two without disrupting the unbroken proceeding. The same with even and odd whole numbers; between 2 and 4 or 3 and 5, no other numbers can be inserted without disrupting the continuity. This can also be done with fractions. A unit can be divided by 2 over and over. Between 1/2 and 1/4, no number can be inserted without rupturing the continuity of division by halves. This kind of continuity contradicts their test.

In order to get around this, they have to define any continuity which cannot pass their test as some kind of discreteness.

Note that the Bolzano-Weierstrass definition is not made positive through the naming of an attribute, but through the statement of a property, i.e., if one performs a certain test, a specific result will take place. This is not wrong, invariably. But for purists who want to exclude time as an extra-logical consideration, this is not consistent thinking; the "if" necessarily precedes a later point in time.

Man's awareness of quantity arises from two sources. One is his experience of the discrete: a small child notices changes, varying colors, moving objects, etc. The other is the experience of the continuous: the same child observes movement without breach from one place to another, differing lengths, balls with the same shape but of various sizes. From the side of the discrete, number arises through counting. From the side of the continuous, number arises from measurement.

When a notion so fundamental as continuity is given a restricted meaning, there is great danger of missing the wider aspects.

Modern mathematicians will answer that all that is necessary is that they be consistent, that there is nothing wrong if they designate the uninterruptible counter-examples mentioned above as mere "discrete series." If they wish to relegate the class of all the natural numbers (or the first n of them) to the discrete, they may.[52]

How different was the practice of the ancient Greek mathematicians who identified numbers with lengths. What we characterize as "1" was to them a standard length, which was to be multiplied to find greater numbers, or divided to form fractions. The combination of that system with the discrete created the number line. Once the unit is defined, it can be multiplied to any length. It can be divided into any rational quantity, any whole number or fraction, and these can be precisely identified. Each would be symbolized with standardized notations, such as 1, 1/2, 1/3, 1/100, etc. Irrational quantities, although they cannot be determined by discrete numbers, must have their place. Once the unit is set, one knows in advance the point which must stand as the endpoint—say 2—for the interval between 0 and itself, 2. By necessary implication, any number, rational or irrational between these endpoints, must have a position on that line, even though a person may never be able to discern which point corresponds to such irrational magnitudes as the square root of two or three. And, of course, the line may be extended, indefinitely.

But the idea of the last century and a quarter is very different. The number system is no longer to be based upon a spacial idea of a line. Instead it is to be a purely abstract creation of the human intellect. Dedekind and Cantor claimed that it would be possible to define abstractly a continuous number system, even if geometrical space were not continuous.[53] The 20th century philosopher Bertrand Russell said that the homogeneity of space is only empirical[54], by which he meant, of course, that we might later on find a part which is discontinuous.

The modern notion of continuity, already criticized here, fits in perfectly with this idea, since it supposes an infinite division of interval within interval without necessarily supposing a line with points. As Carl Boyer put it, "The mathematical theory of continuity is based, not on intuition, but on the logically developed theories of number and sets of points."[55] For them, continuity is the sequence of digits, some of them extending to infinity—as if infinity were a stop.

Let us accept the challenge that mathematical continuity can exist apart from the number line—in fact, apart from spacial considerations. It will be shown that what is conjured up according to the modern theory of continuity is not really continuous; that it is really inferior to the number line which is based upon a spacial idea.

Suppose, for the sake of the argument, it would be possible to abstract the idea of mathematical continuity from space. Even so, one could not eliminate time. The if-for any-number-there-exists-a-number smaller idea presupposes simultaneity. If it were not simultaneous, then when one looked for that smaller number, one might not find it because it was not there at the time that the selection was being made.

Having shown that they cannot escape from time, let us return to the spacial question.

But if a sequence of consecutive integers is to be classified as a discrete series when there are interruptions, albeit regular, why must the rational and irrational numbers in the sequence between say 1 and 2 be necessarily continuous? If we are to follow Dedekind and break with geometrical intuition, how can one know that our free constructs will always yield that which is required by the modern definition of continuity? Perhaps there are unseen and unforeseeable breaks?

In the 1930's, one of their own people, a man named Gödel, argued that this was exactly what they could not prove; they could not show that within their premises, the number system they were building was consistent. To do that, they would have to go outside their number system. His idea has since become orthodoxy among them.

Gödel's thesis unintentionally presumes time also, since it teaches that the theorist who attempted to be completely consistent in his free construction could later on find that it contained propositions inconsistent with his original premises. If the theorist then tried to correct it by adding a premise to make the whole consistent, he might then find that the revised system was again inconsistent. It also assumes memory, for if the free constructionist forgot what he was doing, he would never know whether he had encountered an inconsistency. To reply that whether they remembered correctly or not is irrelevant—that what matters is the truth—is to recognize that free constructions proposed by people like Dedekind are radically insufficient. And even if that is granted, how can one know that the subjectivity of another might be different?

Against this, it might be objected, how can we know that the number line might enter a realm so contorted that continuity would be broken?

To that, two answers can be made: (1) the frivolous one that in that case, which would be impossible anyway, one would simply go back to the length that made sense and re-calibrate, using a smaller unit. If they are right in thinking that any interval can be divided in an unlimited way, a mathematician could simply extend the line into whole numbers it had previously not reached by applying shorter units to the lengths that had already been tested. (2) The serious answer that it is not reasonable to count something as a possibility simply because one can imagine it. In the words of a leader of the Objectivist school, Leonard Peikoff, such an argument "confuses Walt Disney with metaphysics. That a man can project an image or draw an animated cartoon at variance with the facts of reality, does not alter the facts; it does not alter the nature of the potentialities of the entities which exist."[56] (Gödel's argument will be discussed once again in Chapter XXI).

To return to the main point: the heirs to the school being criticized have concluded that the path charted by the free constructs is confronted by the endless task of never being able to justify itself. This being stated, let us return to the discarded intuition based upon the line and see what is missing in the modern notion of continuity.[57] In so doing, we shall also show why the old number line is inherently superior to the alleged continuity built up by these intellectuals.

Clarifying Dedekind and Cantor, Edward Huntington divides every mathematical series into three categories, discrete, dense, and continuous. In a discrete series, every element, unless it be the last, has an immediate successor and every element, unless it be the first, has an immediate predecessor.[58] Huntington divides discrete series into four types, one type of which has a first element but no last; these, he calls "progressions." An example is the series of natural numbers $1, 2, 3, \ldots$. The first element is 1. (These correspond to Cantor's type ω.)[59]

The second basic division is the dense series, typified by the rational numbers—Cantor's type η. Common to it and the third division is the density postulate, which is most important for our discussion.

The third division is the continuous series, typified by the real numbers, which includes both the rational and irrational numbers.

The Postulate of Density is: *"If a and b are elements of the class K, and a < b, then there is at least one element x in K such that a < x and x < b."*[60] This Postulate is crucial to the modern definition of continuity, for that too holds that between any two members of a continuous set, there must be another element.

The modernists differ as to whether the real numbers have a first element. Quite often, zero is not called a first element, in which case there is no first element. Both possibilities will be covered in this chapter.

The number line contradicts the density postulate. If there are no points next to each other, there is no actual continuity. A broken connection anywhere is something less than that, i.e., a discontinuity. Without consecutive points, there could be no line. Therefore, under the modernist premises, not only do they present an alternative to the line, but they deny the possibility of a line, even in principle.

Their continuity is never absolute, but limited only to the actual finding of a number between two given numbers. If one could always find points between any two points in a line, then every part of a line would be under construction all the time. What allows them to assert the postulate of density with confidence is the fact that no one has yet found numbers below the level of the unit which succeed each other; but this, as we have seen, comes from the fundamental immeasurability of the level of the infinitesimal. Furthermore, although they might wish to expunge time from their theoretical constructions, they cannot.

They also depend upon the fact that everyone who reads their declarations already has a notion of continuity. Yet, their position is fundamentally contradictory, since they base their test upon numbers, which are notions which are impossible in the absence of the more fundamental ideas of continuity and discreteness. This is an instance of the fallacy which Ayn Rand rightly identified as the "stolen concept."[61] They attempt to invalidate the natural understanding of continuity and discretion by using numbers, which are derivative concepts.

Why is the number line free from contradiction? The reason is that the points in a line are only locations. Two or more cannot occupy the same space, for each is its own space. These points are not round; they have no shape, as form is a finite attribute. And the points are infinitesimals. They are not intervals, for the two endpoints define the interval.

Would the modernists be right if they reduced their definition of continuity to the level of being no more than a species of the general concept? In order to answer that question, let us begin by counting the position zero as an element of the line.

Take the point just above 0 on the positive scale. This point would have to exist. If it did not, there would be no continuity at that point, which is absurd. The modernists divide real numbers into rational and irrational numbers. If this point above zero were a number, how would it be characterized in decimal form? (. . . 0001) is what it would have to be. The ellipsis would signify an endless number of zeros, followed by a one. Now, suppose, *impossible*, that such a number could exist. Would it be an irrational number? Square roots are operations upon an actual number; for example, $\sqrt{2}$ is a number which when multiplied by itself, produces 2. Similarly, transcendental numbers, such as π, are the result of operations upon fundamental geometrical figures or the limit of a series, like e. But the point which follows zero is not obtained from any finite number. Suppose it be squared, i.e., multiplied times itself. The resultant would have to be smaller than (. . . 0001), just as the square of .05 is .0025, a much smaller number. But no such product can exist, since by definition the only alternative to the point just above zero would be zero itself, which could not be a square. Yet, this position surely exists. It could not be handled through approximation as are irrational numbers, for the product would still have to be smaller than itself, an evident impossibility. The only point smaller than it would be zero, and this could not be the answer. It, therefore, cannot be either a rational nor irrational number. It is an infinitesimal.

What would the next higher point be? *(. . . 0002)*. And the point after it? *(. . . 0003)*. At what point would it become a rational or irrational number? Obviously, this cannot be represented by decimals, or by any other mode based on finite numbers, not even as an approximation (unlike the irrationals). Decimal representation is adequate only to intervals.

Rational and irrational numbers require finite intervals. The points in sequence above zero but below the level of the finite cannot be defined as numbers. They are not finite intervals.

Yet, these un-numerable positions exist. To deny them is to destroy continuity. There must be a point next-to. The real number system, as the modernists conceive it, may be internally consistent, but this self-consistency does not even cover all of the positions on the Cartesian axis which it is supposed to represent. The same is the case with the imaginary numbers, which are supposed by them to complete the number possibilities.

And so, the concept of continuity, so popular in the last century, actually lacks that attribute; no wonder they trembled before Gödel.

Let us next consider the possibility of not beginning with zero. The aforementioned position just above what would have been zero on the positive line would still exist. It would then be the least element. All the difficulties of taking its square would remain; also those of succeeding points below the level of the finite.

Whether 0 is included or not, the result is the same. The modern system, with its strict division of the real numbers into either rational or irrational numbers, leaves out positions which exist for the number line based upon geometrical considerations. Those recognizing zero as a first element leave holes of discontinuity and so do those who do not recognize it as a first element.

Let these positions on the number line below the level of a finite interval or magnitude be called "*infinitesimal positions*" or "*positional points*."

A similar situation exists with respect to finite magnitudes themselves. A good way to begin the discussion of that is to consider an idea which has been common since Dedekind. This is to represent 1/4 as 0.2499999 . . .[62], instead of .25. If there were an endless quantity of points within an interval, such an expression could not be carried out. And even if we suppose the impossible and imagine that it would, it could only end with the infinitesimals just below .25 on the number line. While it is true that there is no finite difference between the two, the fact remains that they are not identical. The same thing could be said for .2499999 . . . 8 and .249999 . . . 7. These are not identical either with .25 or with .2499999 . . . . The decimal mode of presentation is inappropriate for less than finite intervals.[63]

Let us consider a point just above or below .25 on the number line, i.e., much, much less than a finite interval greater or less than it. The extra infinitesimals are not enough to make a rational difference. Since its difference is less than finite, it is not commensurate with any rational number.

That this point must exist is obvious; the continuity of the number line guarantees it. It must also be the terminal point of a magnitude close to .25. That illustrates the difference between the rational and the irrational in mathematics.

Why is it optional as whether to count zero as a first element or not in a continuous line? The answer is as follows. Zero is a label given to a position. That position is indivisible. It is not a region, for any case of the latter could be divided into points. Once the size of a unit is determined, then the significance of zero, if it is included, becomes established. Let us take the interval between 2 and 3 on a number line. Where does 2 leave off and 3 begin? This question is easily answerable if we recall that numbers existed prior to the discovery of zero; the Greek numbers were lengths. The unit "1" was defined in the absence of "0". It began with the point just above what would later be labeled zero. Far from being trivial, this point which cannot be defined in terms of finite numbers is indispensable to the determination of the unit.

Now, where does 2 leave off and 3 begin? 2 ends just below the infinitesimal labeled 3. It begins at the "2" mark, and 1 begins just above the infinitesimal labeled "0" and ends at the "1" mark. 6 minus 0 equals 6 precisely because 0 was never part of 6 to begin with.

The other positions marked with numbers along an extent are no different. The infinitesimal labeled "7" is no greater than the one marked "0". It has no finite size, either. Apart from the intervals above zero which they terminate, they have no additional meaning.

People who talk about 2, 6, 8, 10 and so forth as a set of even numbers should keep this in mind. "10" is just a mark over an infinitesimal. To speak of it as representing something *totally different* from 8 when the interval it stands for must include the latter can get someone into trouble; the magnitude of TEN is unintelligible apart from that represented as EIGHT. As a means of representation, it is only two units larger.

Dedekind once suggested that if one were to take a line made of only rational numbers and extend it in three dimensions, i.e., R^3, the Euclidean geometer would not notice it.[64] This, of course, is false. A rational magnitude must include irrational magnitudes; the rational magnitude 2 could not exist if it did not include the irrational $\sqrt{2}$, since the latter is approximately 1.414 units long. Aside from that, one needs only to recall how irrational numbers were discovered by Greek geometers in the first place. Book X of Euclid's *Elements,* which concerns incommensurable magnitudes, is also the longest of the 13 books.

Clarity is required, whether speaking of the infinitesimal symbolizing quantity or the quantity itself. For example, $x \rightarrow 0$ can either mean that the variable x is approaching the position marked zero, or that x is ceasing to exist. It might be possible to speak of a group of five neighboring infinitesimals and out of this group concentrate attention on three, leaving two. But, in absolute space, there is no place for zero which is absolute in all respects. There are two kinds of zero.

The first is the result of an arbitrary decision to designate a certain infinitesimal as "0". The other kind of zero is used to represent the deprivation of some thing, some action or some attribute; for instance, two apples minus two apples means no apples. It is impossible to cancel a position in space. A point in empty space is still a position, whether or not anyone has located it in some system of coordinates.

In summary, Aristotle affirmed that only potential infinity could exist; the Greeks were afraid of infinity. On this last, we find ourselves in agreement with Georg Cantor, who knew that actual infinity had to exist.

But the modernist idea of continuity excludes the next-to. By doing so, they place their continuous lines as ideals outside of space. Since any physical body must exist in space, their ideal is incompatible with matter, light, and energy.

The modernist idea of continuity is backwards. The proof of continuity should be that nothing could be inserted between two neighboring numbers, not that it always can be. The continuity of the number system is not the same as always keeping busy.

Their beloved digital representation, augmented by ellipses, cannot represent recurring fractions and irrational magnitudes. Furthermore, infinitesimal positions exist which are too small for any determination on the part of the unit; they cannot be reached through irrational magnitudes, either. There also exist transcendental magnitudes which cannot be reached by any combination of rational and common irrational magnitudes, even though they are finite.

Man's realm of natural knowledge is the finite; yet, he cannot go far without making use of that which is beneath it; nor can he calculate long without becoming aware of that which is greater than any magnitude. We can suppose a line, which Euclid correctly defined as a breadthless length[65] , going on forever; but that is only a potential infinity. It is no longer than it is at a given moment. Our number system is rooted in the finite and can never extend through unassisted reason beyond that. Man is not the measure, but he is a measurer.

CHAPTER V

THE CONFLATION OF ACTUAL AND

POTENTIAL INFINITY

The school of thought dominant in the 20th century accepted the existence of the actual infinite, but rejected the distinction between potential and actual infinity. They did not stop with the fact that a potential infinity implies that the idea of actual infinity is valid, as was argued in the third chapter. They went in a contrary way and reified potential infinity, contending that the very content of potential infinity was actualized. For instance, they would say that even though the total number of whole numbers (natural numbers or integers) known is always finite, the fact that the process of adding to them has no end means that their number is really infinite. And finally, that they exist in a set.

Advocates of the finite, like Pisaturo and Marcus, counter that numbers in their modern usage are *"adjectives* describing groups, not distinct entities. It is always so much or so many of such . . . . A number, whether in the old 'noun form' or the more modern 'adjective form,' refers to a group, not to some individual object apart from any group. (If this point is understood, then the problem of infinity, which arises from the attempt to reify infinitely-many objects, does not arise. A large number has an actual referent only if there is a group with that many units in it.)"[66] Their objection is that since the number of whole numbers we have on hand (the group) is finite, it is an illegitimate process to extend it beyond what it is part of—the group. In school, we may practice adding, subtracting, multiplying numbers without specifying what they stand for, but these abstractions cannot stand by themselves. Numbers, they conclude, "specify the measurement of the quantity attribute of groups of units."[67]

We agree in part, but our foundation is the number line, rather than some group theory. Since the process by which men increase the number line is in principle unfinishable and is finite at any state of its advancement, it cannot reach a "point of infinity" anyway.

Does this mean that there cannot be such a thing as an infinite line? One can imagine oneself pointing in some direction and proclaiming, "a line from the end of my finger that would not wrap around the earth, but go on forever, never bending, never curving, but always extending beyond even the far stars discerned by the astronomers, never stopping." But there is a great difference between imagining something and actually constructing it.

Where is the difficulty?

A line is sometimes defined as a moving point, which Aristotle probably accepted.[68] This definition is not strictly true for the infinitesimals we have been discussing. No spacial position can get up and go to someplace else. One cannot even define it as an orientation of points, for a direction is an attribute of a line, and wherever a direction is determined, there must be a line.

A line is an expression of motion; it is the tracing of that motion across certain positions in space. When a body moves, a point can be found within its mass that will travel with it, occupying position after position. The path of this finite displacement is a line. In that respect only can one think of a line as a moving point. Since lines are the result of motion, they are as limited as that particular motion itself. Men can imagine a space ship going on forever, but unless they are on it, conscious of what is going on, their lines do not. Yet, for the same reason that we know that infinity can never come within our compass, we also know that actual infinity exists.

Some may say that this conclusion, although reasonable, is not in harmony with that of Euclid whose lines, they think, were infinite. Apparently, the famous 20th century German mathematician David Hilbert thought that this was the case, based upon his trendy revision of Euclid. According to Morris Kline, Hilbert's "Axioms 11_1 and 11_2 amount to making the line infinite."[69] That may be the case with Hilbert, but his axioms are not Euclid's; they are merely an attempt to reinterpret Euclidean geometry in a way that is consistent with the preferred lines of thought of the age of which he was a leader.[1] Sir Thomas Heath, the translator of the standard English version of Euclid's *Elements*, did not think that the actual geometry of Euclid required actually infinite lines.[70] (In the next chapter, Hilbert's work will be brought up again.)

---

[1] Hilbert's Axioms 11_1 reads, "If a point B lies between point A and C then, A, B, and C are three different points on one line and B also lies between C and A." and 11_2 reads: "To any two points A and B there is at least one point B on the line AC such that C lies between A and B." The second axiom, especially, seems to require an actual infinity. But these are not Euclid's.

The distinction between potential and actual infinity is crucial. An example of what happens when the distinction is forgotten is the statement of Richard Dedekind that with infinity, a part of a set is the same as the set itself.[71] "A system S is said to be *infinite* when it is similar to a proper part of itself; in the contrary case, S is to be a finite system," he affirmed.[72] Dedekind and many others have supposed that if one were to take part of what they call a "set"—for instance, whole numbers—and put them in one-to-one correspondence with the entire rational number system which contains both whole numbers *and* fractions, one would never be able to find a whole number beside which one could not place a fraction, and, therefore, that the part was as great as the whole. But this confuses actual with potential infinity. Dedekind was preceded in this by Bolzano, who once more, led the way in the 19th century, although the great Galileo had once made a similar mistake.[73]

Obviously, there is something wrong with this idea. Let it be assumed for the sake of the argument that the points within a unit are infinite in the sense of being endless: suppose a person takes an inch from one English foot ruler and attempts to place it in 1-1 correspondence with all twelve inches of a second foot ruler, figuring that this can be done because the points within the inch are infinite and so are the points within the foot. Assuming for the sake of the argument that a matching of infinities is possible, this person would find out that he had exhausted all of his points when the points on the first inch had been matched with the 1st inch on the second. This fact cannot be ascertained by counting on a one-to-one or any other basis, but by comparing the two equal finite intervals. The end points of the solitary inch and 1/12 of a foot must match. So also must be the possible rational subdivisions and irrational parts within them.

Instead of comparing a couple of material rulers, consider two number lines, A and B. Let us match the even numbers on A to the total natural numbers, 1,2,3, etc. on B. According to Dedekind's theory, there would be a 1-to-1 correspondence between the even numbers on pole A and the even plus odd numbers on pole B.

He was wrong. It is true that a person could go on matching them as long as he wishes, but once he stopped, he would find that the top value on pole A was about twice as large as the top value of pole B. The part is not equal to the whole. Numbers are human creations. The number line itself is finite, although not in principle finishable.

Even if we suppose for the sake of the argument something that is not true, i.e., that the line is actually infinite, the infinite line of natural numbers would be uncountable and so would be the even numbers. 1-1 correspondence is a concept which pertains to finite numbers, not to infinities.

Edward V. Huntington offered as an example of an infinite class, "the class of the natural numbers, since it can be put in one-to-one correspondence with the class of the even numbers, which is only a part of itself."[74] But when one thinks of it, this idea is impossible in that way. It treats these magnitudes simply as points, ignoring that the smaller numbers being indicated on a point-to-point basis are implicitly contained in the larger ones, for instance, the magnitude represented by 5 already contains the even number 4, and so on. Of course, for a "five," there must be "four," because otherwise that quantity could not exist. How would the natural numbers be paired with the even numbers? If 1 were to be paired with 2, then 3 were to be paired with 4, a person would actually be only pairing the odd numbers with the even. To avoid this, this person would have to pair some of the same numbers twice, i.e., 1 with 2, 2 with 4, 3 with 6, 4 with 8, 5 with 10, 6 with 12, etc. But that would destroy the desired 1-to1, for even in the small sample just listed, 2 has been paired with 4, and 4 has been paired with 8.

These are supposed to be numbers in complete abstraction. Yet, what is compared are two distinct number lines, separated in space. Only by considering a "2" from the even pole to be different from a "2" of the natural number pole can contradiction be avoided. In that case, they would be different "sets" and therefore the part could not be equal to the whole.

Once more, let us consider this question. Let us take the finite extent from 0 to 1 on the number line. Within it, let us compare the divisions by one half, 1/2, 1/4, 1/8, 1/16, etc. and the divisions by one third, 1/3, 1/9, 1/27, 1/81, etc. Obviously, this 1-1 comparison would not be completed. Both are potential infinities. Now, compare the divisions by one half in the interval between 0 and 1 with the same sort of divisions within the interval between 0 and 2. Dedekind would say that the two would be in one-to-one correspondence, since each is infinite in the same way. But, as we have seen, 1-1 correspondence is finite and cannot exist on the level of their infinite. We can say that the finite unit 5 is five times greater than the single unit, 1. But even if we say that each of these units contains within it an infinity of points, and each is the same, we cannot say without contradiction that the infinity in the solitary unit is equal to the totality within the five units, which is what one-to-one correspondence really means— *equal in number*. Therefore, the actual infinite cannot be defined as a condition in which part of it can be placed in one-to-one correspondence with itself. It is beyond numbers.

With potential infinity, the state of the process at its highest point of advancement is always finite, but the infinity lies in the fact that men cannot complete it. Potential infinity inescapably presupposes time.

It might be objected that there is something strange about the spacial infinity existing within a finite interval. If the points in an inch are unlimited and the points on a second inch are also unlimited, are not the two inches twice as unlimited? But this is a topic that will be discussed later in this work after the appropriate preparation has been made.

Further contradiction can be found in the Cantorian argument that there are no more points in a square formed by the product of two real line segments at right angles to each other than there would be in one of the equal line segments.

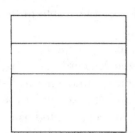

This last is impossible. Let us suppose, for the sake of the argument, that one to one correspondence were possible with unlimited points. Consider a horizontal line segment constituting the lower base of a 1-unit square. Next, consider a line segment of the same magnitude parallel to the base but placed above it by ½ unit. Obviously, every position in this second line segment of one unit would be in one-to-one correspondence with the first horizontal line. Now, let us imagine another segment placed half the remaining distance between the second parallel and the top line of the square. Since the lower base and the first parallel are in one-to-one correspondence with each other, there is no place left for the third parallel line to be in that same kind of correspondence with the lower base; every place has been taken. There is no room for any more. But it is supposed that since the first one is unlimited, there must be some place left for this extra line. This is contradictory. What it calls for is 1-to-1 correspondence, not 2-to-1 correspondence. The answer is that ultimate space is absolute. The first segment is infinite for that particular interval, which must occupy so much space. Even if we move it around, it is at some place or other all the time. The three parallel rows are, each of them, infinite for their segments in their respective positions.

Now, instead of considering three parallel segments of the same length, let us consider parallel lines of the same length vertically stacked upon each other until the square is covered. Here, we clearly see that there could not be any 1-to-1 correspondence, only an infinity-to-1 correspondence. This is what one would conclude if the one-to-one idea could be applied to those infinities.

Georg Cantor argues that c = cc. By c, he meant the power of the real number system. He held that the number of real points in a unit square (c^2) is equal to the number of real points in one of its sides. Actually, his argument was somewhat different from the one refuted in the previous paragraph, although it ultimately rested upon the 1-to-1 idea.

Lillian R. Lieber explains it this way: "Suppose A is a point somewhere within the unit square . . . . This point, A, has two coordinates, x and y, both of these being real numbers having values between 0 and 1. Suppose that x is $0.a_1a_2a_3a_4\ldots$, some real number (rational or irrational) between 0 and 1, and that y is $0.b_1b_2b_3b_4\ldots$, another real number (rational or irrational) between 0 and 1. Now, obviously, from both [of these digitalized numbers] it is possible to form one SINGLE real number thus: $0.a_1b_1a_2b_2\,a_3\,b_3a_4b_4\ldots$ and this resulting number also being between 0 and 1, will be located somewhere on the X-axis between 0 and 1, will it not? And similarly any real number between 0 and 1 on the X-axis, like $0.c_1d_1c_2d_2c_3d_3c_4d_4$ . . . , may be split into TWO numbers, $0.c_1c_2c_3c_4$ . . . . , and $0.d_1d_2d_3d_4$ . . . . , which may serve as the coordinates of some point in the square . . . . And so there is 1-1 correspondence between the points on the LINE-segment on the X-axis, from 0 to 1, and the points in the SQUARE, so that these two sets of points are equivalent."[75]

In the example given by Lieber, the digits represented by $0.a_1a_2a_3a_4$ . . . . could only amount to a length measured from zero on the X-coordinate axis; similarly, $0.b_1b_2b_3b_4$ . . . . could only be a length represented on the Y-coordinate axis. If the first were summarized as α and the second were summarized as β, their point of intersection could only be represented as α, β—never as $\alpha\beta$, which would be the product of the lines formed from the intervals between 0 and α and 0 and β, respectively. Square feet or meters are in principle different from linear feet or meters. (There are additional fallacies, but these, we will not discuss.)

In the last chapter, the futility of representing a number like, for instance, "1" by an infinity of 9's was shown, that $99/100 + 999/1000 + 9999/10000$ may approach one, but cannot ever equal one. But the whole idea that a person can represent a number with an infinite number of digits is fallacious, because numbers can only be finite. Digits are not like locations within an interval on a number line. The locations from 0 to 1 exist simultaneously, whether identified or not. The digits have to be calculated one by one. Even if these calculations proceeded at the speed of light or at whatever is fastest in existence, this would still take time and therefore cannot be indicated with a mere ellipsis.

Even if one takes an approximation which one knows in advance can consist only of a single numeral and requires no additional calculation, the result is the same. Consider .33333 . . . , which is supposed to equal 1/3. The truth is that it is not the same as one third. It is an approximation; it cannot meet its objective. The fact that we know what each succeeding place must be in advance of additional calculation makes no difference. Since there is no precise decimal equivalent to the fraction 1/3, it will never hit the mark. To consider an estimate which can always be improved upon as the same as actual equality is to make a grave mistake. Moreover, the digital mode of representation is not specific enough when applied to infinity. In many cases, it cannot point to an exact number. The first nine digits of the following "number A" : .5432078654 . . . . would be the same as the first nine of "number B": .5432078653 . . . . , but thereafter, they will differ from each other. Therefore, using the first nine digits is not an approximation of A, since they are also those of "B"—not only of "B," but also of those with tenth place digits of 0 or 1or 2 or 5 or 6 or 7 or 8 or 9. In fact, since this multi-ambiguity occurs at each additional place, the confusion progresses. Take, for instance, the "number" which begins as, say, .0000895587321059382 . . . . It might be that the number which the mathematician thought he or she was indicating was, in fact, something else. There might be a different number which contains the same initial 18 digits of the number wanted, but differs on the next one. In fact, by that kind of reckoning, there would be an infinity of other numbers, each one differing from its predecessor by one digit. It is not enough in a case like this that the desired number be approximated within the context of one's knowledge, i.e., so that no confusion could exist with respect to two or more numbers which one knew; it must be absolutely specific. $0.a_1a_2a_3a_4$ . . . . is ambiguous. By contrast, the symbol, $\sqrt{2}$ is completely specific, even if it needs to be approximated in real life problems.

The same is true for a sequence which appears to be a rational number. If the sequence has repeated itself a hundred times, we do not know for certain but on the next try, it will not move into the unchartable.

Even when we ignore the ambiguity and rely on the intention behind such an expression like .0000895587321059382 . . . , the task would still be insurmountable. This is why so many modern mathematicians attempted to abolish the difference between potential and actual infinity. What they meant was what the digits would be if one of their infinite strings were ever completed. But, as was explained above in several places, it is the nature of the situation that it can never be completed, *even in principle*—once they make the assumption of endless infinity.

That being the case, such numbers do not and cannot exist. This was also shown in the last chapter in which the inability to indicate the finite number of points above zero on the number line was discussed.

This fallacy stands behind much of modern mathematics. Although Cantor correctly saw that the notion of potential infinity implied that there must be an actual infinity, he drew the false conclusion that because the irrational numbers can only be approximated in digital form, this means that they somehow exist in an endless digital form.[76]

As an example of how this fallacy can be avoided, consider the famous case of the map of London. In the words of Edward Huntington: "Suppose a complete map of London could be laid out on the pavement of one of the squares of the city; then the city of London would be represented an infinite number of times in this map, and the successive representations would form a progression. For the map itself would form a part of the object which it represents, and would therefore include a miniature representation of itself; this representation being again a complete map of the city would contain a still smaller presentation of itself; and so on, *ad infinitum.*"[77] (This idea was originally suggested by Dedekind.)[78]

Let us leave aside the practical problems of attaining such resolution. Even so, providing that it was the finite dimensions of the city, including the map on that spot of pavement that was being represented, the map would only illustrate familiar features of finite intervals. However, if there were an attempt to represent not only the finite features, however tiny, but also each and every infinitesimal in the features shown on the map on a 1-to-1 basis, there could be no map inside of a map inside of, etc. In that case, the dimensions would have to be the same as the original city, for each infinitesimal in any finite interval of any size would have to have its match. Furthermore, taking the map as it is, if this magical map were completed, then all of this infinitude of representations would be simultaneously present—an actual infinity; and if it were done, one after another, as suggested by the term "progression", then it would never be complete. It would have to be a potential infinity. But then it could never be fully accurate.

In the last chapter, such infinitesimal positions as the infinitesimal just above zero was discussed. The number line has a first element, even when zero is left out. Under the definition of continuity used by the school of thought being examined here, there is no first element—no first element, because no "next to." Between any two numbers, another is supposed to be always awaiting discovery. Not having the distinction between the finite and infinitesimal, this presented this school with a problem. They could not completely achieve their one-one correspondence unless there was a least element.

The class of the natural numbers had a first element, 1, although no last element. This class, which Cantor called type ω, he considered. as well ordered. "By a *well-ordered* set," he explained, "one should understand any well-defined set in which the elements are bound to one another by a precisely given succession, according to which there is a both a *first* element of the set and to every single element (provided it is not the last in the succession) there follows a certain other, as also to every finite or infinite set of elements there corresponds a certain element, which is the *next* following element of all them in the succession (unless there is absolutely no follower to them all in the succession)."[79]

The next type was η, the class of the rational numbers. Given any fractional interval, for instance, 1/4: the Cantorian investigator could always find one smaller, 1/16, 1/32/, 1/n. It was "simply ordered," in that it possessed neither a greatest nor a least element; and between any two elements, there would exist infinitely many other elements.[80]

The problem was to give the set of rational numbers a first element so as to make it well ordered, like the natural numbers. Cantor's famous solution was to put them in an order different from the actual one so that each number does have an immediate successor. The following illustrates his idea:

```
1/1,  →  1/2,   1/3,  →  1/4, . . .
      ↙     ↗    ↙
2/1      2/2,   2/3,   2/4, . . .
↓     ↗    ↙
3/1,    3/2,   3/3,   3/4 . . .
       ↙
4/1, 4/2, 4/3, 4/4, . . .
```

.

.

[81]

The first row consists of all the rational numbers with a numerator of 1 and with denominators of 1,2,3,4, etc. The second row consists of all the rational numbers with a numerator of 2; the third row consists of all the rational numbers with a numerator of 3; the fourth row of all the rational numbers with a numerator of 4. In this way, he hoped to include the infinity of positive rational numbers. As Lieber comments: ". . . it is now quite easy to show that this is a DISCRETE set, for, by following the arrows, we can go from one rational number to the NEXT, since now each one has an immediate successor. And thus we see that the set of all positive rational numbers, though DENSE when arranged in order of increasing magnitude, CAN be RE-ARRANGED so that it can be shown to be a 'countable' or 'denumerable' set, just like the infinite set of integers, with which it can now be put into 1-1 correspondence."[82]

Is this an improvement over the number line? Consider the order of the arrows; the first step is 1/1 or 1; from that one goes to 1/2, which is included already in the unitary 1; next to 2/1, which is twice the former; then to 3/1, which is 3/2 times the former; then to 2/2, which is actually no different from the first step in that both are equal to 1. This portion is not a sequence, since it involves a repetition. The symbolical expression is different in all cases, but it really does not move from one value to a different one, as does the number line. Even if it were possible to exhaust the rational numbers through this technique—which it is not—some values would receive more than one correspondence, thereby preventing a true 1-1 correspondence. What is unique is only the symbols themselves, not what they mean. With the natural numbers with which they are to be paired, each number is unique in value and also successive in value. Neither is the case with this technique. Cantor could not anticipate in advance all that he would need to throw out.

The alternative is to use decimals instead of fractions. This reduces to some extent the problem of using different fractions which express the same values, but not altogether. With it, a new problem is introduced—that of inexactitude. Under Cantor's formulation, for instance, the fraction 1/4 would be represented as both .25 and .2499999 . . . . Those which contain 9's, he threw out. (The fallacy of making this identification was discussed in the previous chapter.)

The system of 1-to-1 correspondence is impossible with infinite strings of numbers; they can never be presented together simultaneously. The natural numbers used to express the digital places are never anything, except potentially infinite. Using ellipses after each row of numbers so arranged does not make any of those rows infinite.

Expressed as digits, rational numbers terminate in a finite number of places, either endlessly repeating the same digit, or repeating endlessly a group of digits. Suppose for the sake of the argument that it were possible to arrange each of these repetitious patterns in terms of increasing length, beginning first with those which repeat a single number, such as .1111 . . . ; next those which repeat two digits, like .757575 . . . ; then those that repeat three digits, like 037037037 . . . . and so on in larger and larger groups, each group one link longer than the one before it. Every group consisting of a specific number of digits would be arranged in terms of increasing magnitude; a repeating group like .027027027 . . . , which approximates 1/37, would come before one like .037037037 . . . , which approximates 1/27—because of the latter's greater magnitude.

Even if we were for the sake of the argument to accept the legitimacy of using the ellipses, a completed list would be impossible. None of these groups within a rational number could contain an endless number of digits, for if one did, it could not be repeated, and would in fact be irrational. So each of these groups would have to contain less than endless elements. But no matter how great the number of digits a member of a repeating group contained, it could be increased. There cannot be a greatest number N, since there will always be an N+1. There is, therefore, no end to digits within this group. Since at any given stage of its advancement it would be finite, enumeration is a process that cannot be completed; and therefore, it would have to be a potential infinity.

As the mathematical reading public knows, when Cantor tried his placement scheme with the real numbers, which includes irrational as well as rational numbers, he found that it would not work. From this, he drew the conclusion that the real numbers were a greater kind of infinity.

Cantor's idea was to first arrange the real numbers from 0 to 1 in their non-existent actual infinity, preferably represented in the digitalized form. He said that you could then make another list which would be different. Since the rational digitalized fractions within the "set" exhibit repeated regularities, all he had to do to spoil the pattern was change one digit in each of them, thereby creating another list. If the 3 in the second repetition of the rational number .037037037 . . . . were changed to a 4, the irrational number .037037047037037 . . . . would be created. Since he could transform the "infinite" table of rational numbers as often as he wished, the conclusion was drawn that the real numbers were of an order of infinity higher than that of rational numbers. He also thought he could do this with the irrational numbers in the original list of real numbers, which issues in even more problems. But let us not go into that.

On the surface, his argument seems to be plausible. Even if we remember that the . . . cannot betoken actual infinity on the level of the approximates, one can see that just by using this same sequence, he can the next time take the third repetition to replace the 3 with a 4 or any other number, thus .037037037047037037 . . . . He could make irrationals out of rationals to his heart's content.

Does the conclusion that the irrational numbers are greater than the rational numbers follow from his argument? It has already been shown that the digitalized rational numbers are finite at every step of the process, their infinity being only potential. The same would have to be the case for the digitalized irrational numbers. They are potentially infinite only.

But, even on a finite level, are they really greater? Let us take the transcendental magnitude e, which is approximately, 2.71828 18284 59045 2336 02874 . . . .[83] Let us make a rational number out of it. First, let us decide to make the repeating pattern, the 1st nine numbers after the decimal point: 2.718281828. Then our new rational number is approximately, 2.718281828 718281828 718281828 . . . . Of the 26 digits first listed above as an approximation for the point e, 10 were unchanged and 16 were changed. As a person goes through the levels of approximation, all he has to do to convert a sequence of digits presented as betokening an irrational number to a rational number is to arbitrarily order that the others be changed, to whatever degree of refinement the process of calculation has advanced. The 1st 91 digits of the new rational number, e', will be 2.718281828 718281828 718281828 718281828 718281828 718281828 718281828 718281828 718281828 718281828. Suppose the Cantorian tried to best this by bringing forth the first 101 digits of e, namely, 2.71828 18284 59045 2336 02874 71352 66249 77572 47093 69995 95749 66967 62772 40766 30353 54759 45713 82178 52516 64274.[84] The advocate of the rational could trump that by making it the first hundred and one digits of e'', which would thereafter endlessly repeat itself.

If we accept the Cantorian premise that the potential is the actual, the transformation of any irrational number into a rational number can be accomplished simply by inventing a rational number which has in its cycle the same numbers reached by the calculation of the subject irrational number—a number which must always be unequivocally finite—and Voila! the substitution can be made. The choice of how many numerals to work with is arbitrary; instead of the first 25 numerals of e after the decimal point, the first 250 would be OK.

Suppose the Cantorian gives up the claim to have drawn up a complete list of rational numbers. Even on a finite example by example basis in which there are no major orders given to the number system, for every example offered by the Cantorian of a rational number transmogrified into an irrational number, the same can be matched with the opposite example. Returning to the illustration with e, any additional calculation the Cantorian will choose to make on e to restore its irrationality after the first repetition of the original sequence will be matched by the practitioners of the e' and its successors. It is true that in making this conversion, a much greater number of changes is needed for e'-types, while only a small number of changes is needed for the process described by any change made by the Cantorian in order to produce an irrational out of a rational number. But it still remains the case that for every new irrational that can be created out of a rational, a rational can be created out of an irrational, although admittedly much more laboriously. What one can say is that the point e is not far from the one herein labeled as e'.

It should also be recalled that near e are certain points which in conjunction with 0 or the beginning, make possible irrational magnitudes which differ from it by less than a finite magnitude. This would also be the case of any powers of e, such as e^2. Near it would be the second power of these irrational magnitudes. The same would be the case with $\sqrt{e}$. This would also be the case with the e'. Near it would be some irrational magnitudes. They would differ from e' by a finite number of infinitesimals. Operations with their powers and radicals would exist.

No real number can exist unless it has first been identified. What exists are the points. Numbers, to exist, must be established as quantitative relationships among and between points separated by intervals, according to some standard that serves as a unit. Without this act of discovery based upon the decision as to what the unit is to be and where to set it, there are mere possibilities. Since numbers are human creations based upon objective reality, they cannot be said to exist merely because we can imagine that more of them will be invented in the future. (As will be explained in Chapter XXIII, the future does not exist at present. Neither does the past).

For the reasons given, Cantor's proof is not solid; it relies on the permanently unfinishable. However, this does not mean that he was wrong in thinking that the quantity of irrational magnitudes within a unit length are potentially greater than that of rational magnitudes. His argument is strongly indicative, even if not definitive. It is a reasonable conjecture to suppose that because of the greater ease with which irregularities can be produced in a rational approximate than can regularities be made out of irrational approximates, that this discrepancy points to a fundamental asymmetry.

Beyond that, if the numbers to the right of the decimal point were *not* endless, then his proof would be solid; the capability of the opponent on the side of rational numbers to convert his ellipses into repetitions would eventually end as the possibilities were exhausted. But, as it is, his ideal of a line is outside of space and therefore, of reason.

CHAPTER VI

ON DEFINITIONS AND UNIVERSALS

Now that certain obstacles have been moved away, the road is sufficiently cleared for thematic exposition.

The correlated concepts of discreteness and continuity are the fundamental ideas behind mathematics; the number line preserves and presents both. The discrete and the continuous cannot be defined in terms of numbers. But these two concepts are distinct and recognizable to the mind, none the less. Any person who knows how to turn the pages of a book—even if that person is illiterate—knows what continuity and discreteness are. Such a person sees that each page is different and has two sides, and yet that they are connected in such a way that no matter how often one opens the volume they succeed each other in the same way, or in the reverse.

Beyond that, there are, of course, subordinate issues which need to be clarified, especially, borderline cases.[85] These are cases where both characteristics are present in a significant way. An example of the latter is whether to classify a necklace of beads consisting of two colors, alternating regularly, as a case of the continuous or the discrete. The pattern is continuous, but the physical elements are not. An Objectivist would probably say that Edward Huntington had a right to call that a "discrete series." But one would also be in one's rights to call it a "continuous alternation." The Objectivists say that there is choice involved, but once one has begun a particular system of classifying a borderline case, one must be consistent.

Is making it optional always the best kind of answer? Not always. In mixed cases on the border between the two, an answer which does not classify one in terms of the other would be preferable. In the case at hand, that neckless could be called a "repeated pattern," an answer which leans neither way.

The number line is an important mixture; it is a line (continuous) with identified magnitudes (discrete); moreover, it is open to further identification within its present boundaries (discovery) and is also capable of indefinite extension (progress). In the concept "number line," "number" is the adjective and "line" is the noun. As to which word comes first, there is no option; to speak of a "line number" would be confusing. The subordination is appropriate; it is only by being set in a line that irrational magnitudes can be fully intelligible.

The discrete and the continuous are both forms of quantity. Numbers are representations of discrete elements of quantity, defined by reference to a unit.

The number line is straight. A seeming exception to this would be a cylinder, but once the curve was unwound, the result would be a straight line; the same with a cone. An unstretched circle, although the parts would be interchangeable, would not do because it is closed and not open to further extension; to get around this difficulty, the units would have to be re-calibrated every time the number system was extended. An irregular curve would not work, since even though the lengths marked off on as units could be the same length as those on a corresponding straight line, the units would not be interchangeable, due to the different shapes of the segments.

The typical 20th century answer to this line of reasoning would be that a straight line does not have a clear definition. They cite a certain property of a straight line which is also sometimes mis-identified as a definition; this is that it is the shortest distance between two points. They object that on the surface of a sphere, the shortest line is actually a curve.[86]

No doubt about it, there are no straight lines on the surface of a sphere. The closest one can come to that is a line that does not wobble laterally on the surface. But this does not take away from the fact that the straight line between the same two points on a sphere can still exist, for all one has to do is draw the chord through the sphere to prove it. *The fact that a straight line touches a curve at only one point makes the tangent of utmost importance to the study of curves.* Try to imagine the infinitesimal calculus without it!

These moderns imply that curved lines have better evidence than straight lines. Yet, their actual practice is somewhat different. Since the 16th century, most physicists and mathematicians have preferred Cartesian coordinates over the traditional methods of the ancient geometers. This system defines curves by reference to at least two straight number lines; in this most popular mode of geometrical construction, curves are made intelligible by straight lines. Imagine the conceptual difficulty of turning it around and trying to construct straight lines through curved coordinates! Handling the infinitesimals on the level of the finite would be unimaginably difficult.

A materialist who based his geometrical ideas upon the ease or difficulty of physical construction might suppose that there are no real curves in nature. He or she could think that every curve is really composed of tiny little straight lines joined together as in a polygon, these lines so small that most cannot be detected by human art. After all, circles are approximated through such a method. How would one refute such a person? The best way would be through recourse to infinitesimals.

But the typical modern thinker cannot imagine such a turn of the table. Instead, he blithely goes on slighting the straight line. He mocks Euclid's definition of a straight line as a "line which lies evenly with the points on itself."[87] He claims that it is too dependent on sensuous observation, that it is not abstract enough. He wants something more refined.

The typical 20th century refinement was that of the famed German mathematician David Hilbert, who wanted to leave such elements as "point," "line," and "plane" undefined; he also wanted to construct certain axioms which would direct their use. The question of the actual meaning of these elements, he did not regard as essential. He said that far from being necessary truths, they could be replaced by tables, chairs, beer mugs and other objects.[88]

Let us examine the first two of his axioms, the "Axioms of Connection." 1_1 "To each two points A and B there is one line a which lies on A and B." [89] If a point were a chair and a line were a table, this axiom would not always be true. If it were, the table would always have to lie on the chair. If this were reversed and points were tables and chairs were lines, the result would be ludicrous. Obviously, then it does make some difference what is used for points and lines. Consider Hilbert axiom, I_2: "To each two points A and B there is not more than one line which lies on A and B."[90] Inspection will show once more that this would not work with tables, chairs, beer mugs, etc.

Suppose it were the case that more than one sort of existent would fit: in that case, Hilbert's claim to have restated Euclid's geometry in a new form would be false: his restatement being ambiguous, it could as easily represent something else; it would not be unique to it. Furthermore, if it were really true that it was a matter of indifference whether Hilbert's undefined axioms were physical objects like tables and beer mugs or lines in space, then in the course of proving a

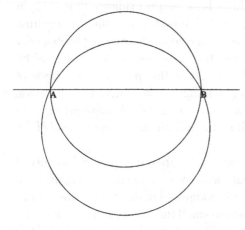

theorem, one should not be surprised if an occasional wooden leg or polished surface came out in the steps.

Actually, Hilbert's axioms depend upon a prior knowledge of geometry. They require that one can distinguish a line running through a point, which is what he means by lying on, from a line lying upon a point. Without this distinction, the second axiom would be false. (In Chapter XVI, we will find that knowledge of this distinction is crucial with respect to the tangent.)

Euclid's *Elements* include circles as well as straight lines. Depending on what part of a circle is cutting a line, more than one can cut through the same two points, A and B *at the same time*. It might work with beer mugs, though, since most of them will not cut through a table.

It is clear from this and Hilbert's remaining axioms of connection that the "lines" he was thinking of were straight. Without a doubt, he had a clear idea of what he was talking about when he composed these and the rest of his axioms. Unless he knew what he was talking about, he would not be able to tell if his axioms were consistent in application or not—or even if there was a realm of existence to which they applied.

Some academician has claimed that Hilbert's axioms cover some of the same material as does Euclid.[91] Assuming for the sake of the argument that this is true, the only way this person could know if Hilbert had been successful in restating some of Euclid's principal ideas would be if he already understood what those ideas which he was restating were—the way they were understood by the original author and his various commentators. If he were unsure of what he was comparing Hilbert's work against, he could not pronounce a sure judgment.

Nor could the ordinary reader have had the foggiest notion of what Hilbert was intending to convey unless he already understood what a line or a point or a plane were. If no one had any idea of what a straight line was, I submit that no one else would be able to discern what was being conveyed.

The undefined ideas Hilbert was working with are thinkable if the reader already knows in advance what the theorist is hinting at, but not if he does not. Such a procedure confuses the act of abstraction, which involves selection, with a bracketing out of knowledge of reality.

Euclid's actual procedure was probably this: He began with certain demonstrations and definitions which had been done by others. After this, he began constructing other figures and reasoned from the rough particular which he had drawn to the general idea of that figure. Then, starting from these, he *induced* the existence of some axioms, which are common notions in terms of which these demonstrations could be made fully intelligible. His axioms were derived from human experience, not from some abstract assertion of the theorist; an approach like that of Dedekind who said, "I require that arithmetic ought to be developed out of itself"[92] was not his. Euclid's emphasis was not on creation, but demonstration of the unforeseen from the obvious.

Only after this did he begin to deduce additional propositions as consequences of these axioms. The idea of writing up fundamental axioms without any idea of what he was talking about, of being concerned only with the goal that they be used consistently, would have seemed frivolous, even disorderly, to him. His science, like all other sciences, proceeded in part through both induction and deduction.

Axioms are not instructions on how to use words we think we don't need to define. Axioms are higher premises upon which lesser premises and theorems depend. It is not enough that they be applied consistently. Consider the following syllogism: *All cows have three horns; this pencil I have in my hand is a cow; therefore, this pencil I have in my hand has three horns.* In this argument, the conclusion follows from the premises in accordance with the axioms of syllogistic reasoning. If it were not so ridiculous, someone might begin by noting that the conclusion is consistent with the existence of both pencils and horns; that since both of them survived the reasoning process, this proves that the pencil in my hand has three horns, even though due to the limitation of human sense organs, it cannot be perceived.

But the truth of the matter is that although the axioms of syllogistic reasoning were observed, both premises were false, ending in an insupportable conclusion. Definitions have to be consistent with both themselves and with what we know of reality.

It is also obvious that Euclid would not have even dreamed of attempting to deduce more than a few of his propositions without using drawings, drawings which would attempt to depict points and lines. This, we can readily "deduce" simply because he also set up postulates—which were practical instructions regarding which constructions were allowed. On this point, John Cook Wilson, a great English logician of the late 19th and early 20th century, once commented. "The real value of the postulates is that they represent an imperfect recognition of the necessity of constructions, that is, of particular figures, either in experience or imagination. They should then be a classification of various kinds of theoretic construction. The classification as given is by no means exhaustive. It does not even exhaust the constructions in the *Elements* of Euclid . . . . But, though it might have been exhaustive of what is actually found in the *Elements*, it could not possibly be complete for geometrical science. The various kinds of construction only become known in the solution of various problems and they cannot possibly be anticipated beforehand. Thus there is no provision in Euclid's *Elements* for the various familiar constructions of conic sections."[93]

(Archimedes worked extensively with the cone and the cylinder. Often, he would produce a conclusion by a method which would obviously *not* be a demonstration, but would give what he described as "a sort of indication that the conclusion is true." Then, he would produce a formal proof; he did not play around with the arbitrary either).[94]

Morris Kline, who admired Hilbert, wrote that "Euclid's definitions of line and other concepts were superfluous. There must be undefined terms in any branch of mathematics, as Aristotle had stressed, and all that one can require of these lines is that they satisfy the axioms."[95]

This is not a good argument. The person using them must have a clear enough idea of them to be able to distinguish them from what they are not. If they were completely arbitrary, they would be unintelligible. Their symbolizations would be meaningless marks on manila paper. Euclid's undefined terms were not blanks or arbitrary propositions. They were terms so basic that they could not be made intelligible by other words, but were found by drawing conclusions from examples from perception, or, perhaps through inborn ideas. When grammar school pupils are given the verb, "to add," the teacher often produces several objects like apples one by one, sometimes varying the illustration until the concept is grasped. The disputed terms of Euclid are identified sufficiently well that the unexpressed part can be referred back to basic human experience. Euclid's definition of a line as "breadthless length." is crucial. His propositions would not be exact, if applied to thick lines. He did not need to define "breadth" and "length." He expected that people would know what all of these words meant by themselves, and that the reader would be intelligent enough to integrate them. The same is the case with his definition of a point as having no parts. The alert reader knows how lacking in superfluity this definition is; why it is important that it be defined right off, rather than forcing the reader to peek around the axioms to determine whether the points have any divisible size, or not.

Euclid defined a straight line as one which "lies evenly with itself." What Euclid stated is a fact of observation. Such a line does not deviate along the line of sight. He did not use the word, "direction," but his meaning is that it has a single direction. A direction is an attribute of a line; a bent line, for instance, has more than one direction, a circle has directions beyond number. Had he used it instead of a description, some critic would have said that his definition was circular; so, he was descriptive instead.

Nobody confuses what Euclid calls a straight line with anything else. Like the discrete and the continuous, straight and curved are too fundamental to be defined clearly in other terms. (It should be mentioned here that the Bolzano-Weierstrass continuity depends upon the performance of an operation, which presumably is to be performed by physical beings, i.e., men.)

By the same token, Hilbert's axioms also rely upon concepts left undefined, besides the words in dispute. Yet, his work is supposed to be above "mere" common sense. How can Kline reconcile this fact with his contention that every axiom system contains undefined terms whose properties specify only the axioms[96], except by relying on common notions, as Hilbert did when he spoke of one thing lying on another?

To return to the straight line, why should we be scandalized by Euclid's definition of a straight line as one which "lies evenly with itself"? As was stated before, what Euclid was conveying is the fact that a straight line has a consistent direction; the concept *direction* is not independent of a line, since direction is an attribute of a line.

Euclid is describing the fundamental attribute that is characteristic of a straight line. It is not necessary that anything found in perception be perfectly straight throughout its extent in order to abstract that attribute; it is enough that the presentation of sense be clear enough for the act of abstraction to take place.

Once one has that attribute in mind—as the Greek master clearly did—one can draw whatever conclusions necessarily follow from it. The inference that the attribute of straightness can be extended indefinitely necessarily holds up as a consequence of its identity. The fundamental experience of forming an abstraction from reality is not a deduction from anything; it is a scientific induction.

Moreover, Euclid's definition is clear enough for subsequent investigators to make improvements upon it. A more scientific definition would be one which connects it with other ideas equally basic. An example is Proclus' refinement: "A straight line is that which represents equal extension with [the distances separating] the points on it."[97] Here, it is connected with equality, extension, points, and distance. Length is not the same thing as a straight line, but if one could not comprehend the latter, the former would also be unintelligible.

That is why defining it as the shortest distance between two points is so incomplete. If, between two points, a loop were placed, of course the length of that loop would be greater than the real distance; anyone who doubts this need only cut the loop and *straighten* it out to discern this fact.

Proclus' definition, like Euclid's, does not separate a straight line from all that it is not in a negative manner, but distinguishes it by a positive property. Like Euclid's, it also connects it with the thought of the infinitesimal. But Proclus surpassed Euclid by identifying it in terms of a property which transcends all lines, straight, bent, or curved, that property being distance. From that, one can readily understand Euclid's attribution that its direction must be single.

This is superior to the definition given in the 19th century by Peano, namely, that the straight line ab is the class of points such that any point x, whose distances from a and b are respectively equal to the distances of the aforementioned class of points from a and b, must be coincident with that class of points.[98] Peano's attempt does connect the straight line with points and the distance between them, but it ignores the fact that the points exist in space anyway and that for a line to exist, it must first be drawn. This was well understood by Euclid; his first postulate was that a straight line can be drawn from any point to any point (at some distance); his second was that this line can be extended. He knew that while no line can exist without points, it is not true that points cannot exist without a line. (Furthermore, the concept of coincidence is not correct here, except as a synonym for "identical.")

Distance itself is a fact known to everybody. It is defined ostensively by pointing out a concrete case to a child who then induces it into a concept.[99] It is so fundamental that great care must be made even in drawing up its properties. For instance, when Bertrand Russell stated that "every pair of points has one and only one distance,"[100] he neglected to consider that two neighboring points have no distance between them.

The discovery of a straight line might have been an induction from a single instance. No comparison between two or more straight lines is necessary; there is no requirement that one first find out if someone else has observed a straight line before one is sure of their existence. The single abstraction of that attribute is sufficient.

What has gotten people to forget that induction from a single instance also exists in mathematics is that so much of the induction of the physical sciences is inconclusive: today's findings may be abridged by tomorrow's, etc. The scientist finds that that A has proved to be B with respect to the varieties of A that he has tested so far. But he cannot say with certainty that all A is B. It may be that there is a type of A which is not a B. He knows that he has not identified all the kinds of A that there are. But he has a belief that when he has explored all sides and aspects of A, he will apprehend the proposition, All A is B, to be true. What he has is a hypothesis; and if it coheres to the rest of his knowledge, he has a theory.

As will be explained in more detail in chapter XII, the difference between induction in the physical sciences and most kinds of mathematics is that the former is so much more complicated. There are so many parts, involving many combinations and permutations, each of which must be investigated. The very means of measurement is subject to perturbations which he may not be able to clearly identify and isolate when they occur. Furthermore, not all the parts are known. The properties of a point and a line are, by contrast, so much simpler. There are unknowns, but they are not the kind which undercut present formulations. The sum of the interior angles of a spherical triangle, such as is found in solid geometry, will be more than 180°, but those of a plane triangle are still equal to two right angles. 1+1 is always 2 in mathematics, but one molecule conjoined to a second one may have some other sum, depending on the circumstances.

In mathematics, a proposition like "All A is B" can be held with certainty when all of the subtypes are accounted for. Even when they are not, the certainty can still hold whenever the connections between A and B are understood well enough to draw correct conclusions. An example of the first is the statement that all ellipses are conic sections; this statement does not mean that an ellipse can only be formed through a cone, but that there exists no ellipse which cannot have been a section of some cone. The proposition is not "analytic" either, but based upon an ability to understand that all possible sections are included in the cone, an understanding that can be had simply by being able to visualize it correctly. A proposition like "all A are B" can also be known with certainty, even when all the subtypes are not known, but the attributes are known well enough to draw necessary connections. A non-trivial example of this would be the statement that all numbers are finite. In this instance, the formation of new numbers could go on forever, but the characteristics of anything which can be called a number is well enough known for its finitude to be asserted.

The chief problem with the word "universal" is the implicit requirement that there be more than one. A "universal" characteristic is supposed to be one which various instances of a certain kind of thing have in common. But a characteristic, if it is actual, should still hold even if only one example of it existed. Suppose, as have many, that there was at one time only one man in the whole world and none other, not even a woman. That being premised: it would still be the case that this man would exhibit the well known characteristics of his species, i.e., the shape, the form, the intelligence, the capacity. Or suppose that it were the case that there were but one dinosaur left after a catastrophe had wiped out the others. Many imagine that this had actually happened. Would not this remaining giant possess for a while all the major characteristic of his species? Suppose first, *impossible*, there was only one straight object: it would still be the case that this singular thing would exhibit the required attribute. If it did not, then it would not qualify as *straight*. From that, we could retain the idea (especially if we gave it a name) even if we never saw another instance. The common expression, "one of a kind," recognizes this fact.

This fact was understood by John Cook Wilson "The term ['universal'] we inherit from Aristotle is seriously misleading; in his own language it is a definition of a thing not by its essence but by an attribute which is not a property of a *given* universal. If we admit universals with only one particular, the designation 'universal' does not even apply to all universals."[101]

What then is a universal? It refers to a type of fact for which there exists or has existed or will exist at least one instance in reality and it is basically repeatable. This usually applies to an attribute or a related group of attributes. It can even refer to a type of which there has been but one instance, providing that the conditions are appropriate. A few years ago, at the bottom of the Mediterranean was found the rusted ruin of an ancient computer, dating from Roman times. It was capable of calculating eclipses and other astronomical conclusions. If it were the only one made and the inventor also went down in the same boat, then it was still an example of the universal, computer, even if another were never made.

There is another type of fact of which there is but one example and which is not repeatable. This type is not a universal. Facts of this second category cannot be discounted. Any moment in past time, in fact, any period of time—even a millennium—is an absolute fact that belongs in the second category. The same with space: although every part of this infinity is completely indistinguishable from any other—there being no center or periphery—each point in it is unique; it is unique even if we could never be able to know whether we had traversed the same point in absolute space again. Consider, for instance, an otherwise matter-less part of space in which a wheel is turning about itself, endlessly. Every turn after the first impulse would be like every other turn except for the moment itself. The hundredth revolution would differ from the hundred thousandth only in that they would not be the same moment and therefore different. So the non-repeatable exists. This could also be the case with matter. Were the atomic hypothesis correct, and matter consists of identical bits of something (even energy puffs), then every one of these would still be other than each other.

A straight line is a universal. It is enough to recognize a single instance of a straight line in order to obtain the abstraction. It may be that a straight thing had been examined—a creased leaf or a folded piece of paper. In such a situation, one observation would be enough to form the abstraction. With respect to simple things, one observation would suffice. How many times was it necessary to see the color pink in order to be able to recognize the fact that one had perceived it before? Once one has an idea of the color of pink, then one can compare it with red and perhaps discern that it is a combination of that second color and white. How many examples of a circle did it take to form the rudiments of that simple idea? Only one. Subsequent observations may show other facts about them, such as that they may be of different sizes. But that first observation may have been enough to fix the idea in the mind.

Euclid's definitions are fundamental facts of reality raised to the level of abstractions and then differentiated from all that they are known not to be by means of at least one distinguishing attribute.

And in connection with measuring and counting, which derive from the fundamental abstractions of continuity and the discrete, the number line can be made to stand out as an idea.

Basic concepts like continuity and the discrete are not fully defined, because they require each other. But they are not blank either; the abstractions are drawn from basic perceptions. The concepts are expressed as negations of each other; both can be traced back to typical experiences. They are both ultimate facts. An infant could form the idea of continuity from almost anything: the ceiling overhead, the solid color of the blanket; the list is long. The idea of the discrete could have originated from the recognition that the portion of food on the plate was smaller than desired, from the confines of the highchair, many things. These two fundamental opposites cannot be spoken of without reference to the other. By characterizing the infinitesimal as an element of continuity, we speak of it as if it were a discrete. But the location, in the case of space, cannot be cut out in any way. The use of the discrete to characterize continuity is necessary in order to show the uniqueness of each location. But it exists as an entirety, and the knowledge that it exists allows us to use a discrete term, such as "infinitesimal." The nearest conception parallel to it on the side of the discrete is the interval, which implies within this differentiation some sort of sameness. This borrowing of a description from something radically different is also found in the characterization of sensuous perception, in which references can be made to the tone of a color or the texture of a musical note.

Continuity and the discrete are fundamental. Given the drift of the last several chapters, one might be tempted to define "continuity in space" as something like this: "a process, whereby, given a certain infinitesimal in a sequence, provided that it is not a boundary, the next member of the sequence is a neighbor." But this definition is completely dependent upon such concepts as "sequence," "boundary," "neighbor," and the like, which would be unintelligible if one did not know what an existent which followed from another without a breach was. Also, the concept of the infinitesimal could not be reached without a prior understanding of the discrete and its opposite.

A pair of opposites more fundamental than the discrete and the continuous is quality and quantity. Without quality, induction and deduction would be impossible; and without quantity, there could be no discrete or continuous—in fact, no numbers. The only concept more basic than quantity and quality is *Fact*.

Discrete and continuous can each be divided into finite and infinite. If one attempted to define the "finite" without reference to some sort of limitation, one might say something like this: "anything exhibiting somewhere properties of shape and / or size." But if that was all that one knew, one would fail to grasp the significance of this definition, unless one had in the back of one's mind the idea of something that in some way did not have those attributes. This is also the case with the more precise definition of finite as that which is in principle capable of being counted or measured—including state-of-art approximations—by human beings. Finite is really a term of distinction. A unit is not simply a string of infinitesimals; unlike the latter, fractions alone fail to exhaust its divisibility.

We become aware of the infinitesimal of space through contemplating the meaning of continuity. The fact that it is for us an abstraction does not make it any the less factual. The infinitesimals of space follow from each other without a break by necessity. Neither do they have any shape. They are locations, but not areas, for the latter imply more than one point. Each location is different, but they cannot be moved anywhere else. Nor do they interact.

Yet, the mind can abstract this idea, though the infinitesimal can never be perceived, even with a forced atom microscope. Expressing continuity through the language of the opposing idea of the discrete, as if they were independent, allows one to focus upon what would be inexpressible, otherwise. They make the geometrical line possible. Right after stating his first definition in Book I that a point is that which has no part, Euclid defines a line as breadthless length.

The number line must be conceived of as Euclidean (without admitting the existence of any other kind of space). It is a potential infinity. This is not the place to consider whether the non-Euclidean varieties are intelligible or not. Suffice it to say that the units in the number line must be of the same magnitude, must be interchangeable, as Pisaturo and Marcus have pointed out, and must also be capable of being extended without limit. If some reader thinks he or she knows any of these "geometries" that might fit, that person can follow along using that, if it can be done. But the burden is entirely on them, for they must show that this non-apparent notion is cognizable.

CHAPTER VII

POINTS AND INTERVALS

The challenge to the case for the infinitesimal is not simply that it have no contradictions in it; then it would be no more than a clear idea. Even if all the Walt Disney movie cartoons about Mickey Mouse were internally consistent, their representations would still be false. The modernist equation of existence with non-contradictoriness must be rejected outright; identity and consistency are not the same.

The greatest attempt to deny it was Immanuel Kant's *Critique of Pure Reason*. He thought that there were certain ideas of reason that were unresolvable, that the structure of reason is such that as good a case could be made against one of these fundamental ideas as for it; that they are undecide-able. The idea nearest to ours was what he called the second "antinomy of pure reason." The thesis he wished to counter was that "every compound substance in the world consists of simple parts, and nothing exists anywhere but the simple, or what is composed of it." The antithesis was its logical denial. In defense of the antithesis, he wrote: "Now the absolutely primary parts of every compound are simple. It follows therefore that the simple occupies a space. But as everything real, which occupies a space, contains a manifold, the parts of which are by the side of each other, and which therefore is compounded, and, as a real compound compounded not of accidents (for these could not exist by the side of each other without a substance), but of substances, it would follow that the simple is a substantial compound, which is self-contradictory."[102] It is obvious that his anti-thesis fails when applied against the infinitesimal. It is not and cannot be a manifold. It doesn't occupy space; it is a spacial location.

Nor is it enough that its rivals have been refuted, whenever they have made their appearance. That would not establish it. It would then be nothing more than a currently winning hypothesis which could hold its crown only until it was refuted.

The argument of this chapter is that the infinitesimal is a requirement of thought, that anyone who attempts to refute it will be convicted of a *reductio ad absurdum*. That type of argument was used frequently by Euclid, and even more by Archimedes.

There is a common mistake about this that needs to be corrected. The *reductio* does not mean that we can actually conceive something which is later shown to be false. Rather, it is a hypothetical argument in which we are to inquire into what would be the case *if* it were true. Consider the following illustration: If A is B, then C is D. The hypothetical conclusion, C is D is shown to be false. In which case, A is B is inconceivable—inconceivable because it required the false C is D. That is why the argument is so decisive. It is not just the rejection of a rival on the basis of a preponderance of evidence. What is rejected is its possibility of being.

As John Cook Wilson put it: ". . . it is not true that we begin by *conceiving* what we start with in the *reductio ad absurdum* (or in any hypothetical argument), and then find that it is not true; for if ever conceivable it would be always so. In that case the proof would be that it was not true though conceivable; whereas . . . the proof lies in showing it inconceivable and altogether impossible for thought. On the other hand, if the thing was conceivable, it would be true, for it would be seen to be necessary—necessary to thought in the sense of the apprehension of a necessity in the object—and there could be no thought of testing it by arguing from it hypothetically."[103]

This is the standard. If the denial of the infinitesimal should be directly contradictory, or result in palpable falsehoods, then its existence is proven. Let us begin by recalling an argument presented in Chapter III. At both ends of an interval is a point. If there were an interval within one of these endpoints, that sub-interval would have two sub-endpoints. If these were in turn divided, and the results of that division also divided, *ad infinitum*, there would be no definite size to any interval. If intervals had no endpoints, there could be no edges either. If no edges, there could not even be a distinct unit length, for no length could be the same as any other. There would be no determinate size to anything whatsoever.

With no determinate sizes, the difference between rational and irrational magnitudes would disappear; the repeating decimals of a rational number would not retain their precise identities, but would surreptitiously metamorphose into unidentifiable irregularities. A lunatic's surmise would make better sense.

Let us examine the situation from outside the point. A region is an area. An area has parts. A point has no parts. It, therefore, has no area. It is just a location; and if it has parts, then the parts would constitute the locations. The process must stop.

If it did not stop, there would be the equivalent of infinite division. The area would shrink forever. Were that the case, then this would mean that there would be a hole in the shape of the point's perimeter which would grow step by step as the diminishing point diminished.

But at any stage, the vacated perimeter would also occupy space; all along its extent, shrinking points would be required. The space would not even be regular with equidistant parts from a center. There could be no center, for if there were, it would be an indivisible point. That being the case, there would have to be more than one shrinking center. And these would leave their own holes, which in turn would mean more diminishing points. In place of the center of each one of these expanding vacancies would be other points. And the same also for every other point. Since the rims would intersect, the points would also coalesce. This would mean one point everywhere. But a point which extended everywhere would be a contradiction. Therefore, points without parts, i.e., infinitesimals, exist.

Stated differently: if a location were finite, then it would be divisible into smaller locations. If these were finite, then they would each of them be divisible into others. There would be an abyss at every sub-point; and within those, more, *ad infinitum*. In this chaos, no location would ever be just itself, but would be other locations as well. Therefore, there would be no locations at all—an evident self-contradiction.

Consider it in the opposite way: instead of constricting one's gaze, expand it outward. Then, there is no place for the rim of any hole, however great. The whole of reality is one great point in which there can be no filling—actually one great hole. Not even empty space—just infinite evacuation.

There is a second contradiction. Keep in mind that a location is simply space and not an object, such as a black spot. What is required for this denial of the infinitesimal is that the vacuum itself be emptied, which is impossible. One cannot evacuate evacuation.

It is because of the nature of the facts of location and area that this conclusion is reached. As such, it is a logical result of straightforward contemplation of the presentations of the sense system. The skeptic who wants to doubt this cannot do so without denying the efficacy of the senses, altogether. *They would have to quibble as to whether locations, edges, and perimeters presented by such experience really exist.* But if they wish to cast such doubts, how do they know that they have read my text correctly? Edges and perimeters figure in that, do they not? Perhaps what they have read is actually something quite different. Locke once asked of the skeptic of knowledge, why doesn't he doubt memory? If a person were willing to really do that—really put himself into such a state, he would sink below the level of an animal. But if he is going to be a real doubter, why not go beyond that warning?

The denial of the efficacy of the senses with respect to boundaries, edges, perimeters and the like is the inconceivable "C is D" spoken of by John Cook Wilson. If the senses fundamentally lie, so do the readings on the vaunted instruments so beloved by the empirical scientist; even the construction of these machines ultimately rests upon the veracity of human perception in these respects. The symbols so cherished by the algebraists would lose their stability, since they depend both upon memory and an underlying reality which does not change with human wishes, presenting either a different text or shifting meanings. What that would have to be is a world so different from that declared by the senses— so far beyond the philosophical distinction between the primary and secondary qualities—that even using the term "world" would be presumptuous.

Stated differently: Consciousness has a nature, an identity; and its identity is the ability to apprehend reality. The senses are an important source of knowledge. If edges and areas do not exist, then every reference to them is false. The reader is challenged to try to live in a waking reality for a single twenty-four hour period without relying on them once.

Those who take advantage of the inherent incompleteness of experience sometimes proclaim that there must be a microscopic infinity to balance space, the latter of which they call the "macroscopic." They hold that it has to have its complement. They say that since the atom turned out to be composed of protons and electrons and since some physicists have argued that these are in turn composed of smaller particles, the process must be endless.[104] Theirs is the idea behind dialectical materialism, the official philosophy of the Soviet Union.

Macroscopic infinity is easy to recognize, as it a logical extension of the world we experience, daily. But microscopic infinity is difficult even to envision.

Such a theory lacks clarity on the level of geometry. There would be no lines, no lengths, because within any line would be more lines, and within any length there would be a fundamental indistinctness. This would apply not only to the material representations, but also to the ideal lines which such writers think exist only in imagination. They, too, would have their internal divisions. Corresponding to the chaos without would be confusion on the inside—not only from the data coming in, but, to use contemporary language, the in-built fracturing of the processor as well. A thought in the mind would also have to be infinite in depth; it could not be something distinct in place. There would not be any science at all, only art. But it would not be the art of Praxiteles or Michelangelo, but of the artful dodger.

Suppose the advocate of microscopic infinity answers that what he hypothecates does not interfere with the macroscopic, that the realm within the point does not blur outward, that what we in the macroscopic realm call an individual point divides it cleanly from the microscopic realm. On the one side of the point everything gets bigger and bigger, while on the other side, infinitely smaller.

The answer is four-fold: (1) Under this hypothesis, the infinitesimal would still be the final stop for what such theorists regard as merely the macroscopic universe. If they did not affirm the finality of the infinitesimal for this side of reality, they would get lost once more in the chaos brought about by denying its incorruptibility. (2) They argue that without an infinity on the inside of the point, the outer infinity would be unbalanced. What we are presented is not even an image of an hourglass with the infinitesimal communicating between two vast reservoirs opening up forever on their respective sides. Consider: on the supposed macroscopic universe, there are points everywhere. If one followed this thinking, one would have to suppose that there was a separate universe within each and every point on the macroscopic side. An infinity of microscopic realms and one lonely macroscopic realm. But physical balance has to do with an equilibrium of forces, bringing about a cessation of motion. Theirs is an aesthetic analogy which contradicts itself. It is an arbitrary mixture of images without an inner coherence. (3) There is no evidence for it. It is nothing more than an arbitrary idea. It is compelling only to those who want to escape from the necessity of recognizing absolutes without affirming that there is no edge anywhere, that all is a blur. (4) Most compelling of all, the microscopic infinity in which everything gets progressively smaller ends up destroying space just as much as did the version in which there is no clear difference between it and macroscopic infinity.

Some may think that they can avoid such discussion by recourse to what they call "experiment." But, as was stated above, the machinery they construct for this purpose could not be built without the uses of their senses. The same for the pointers or dials or computer readouts. If the senses are basically unreliable, so are their experimental findings. Moreover, the experimenters often make reference to earlier experiments. To do this, they have to rely on the faculty of memory. Furthermore, all too frequently, they also refer to theory, which will tell them whether the results are significant or not and, in advanced cases, even what they mean. They must rely to some extent on reason, the very faculty they are trying to minimize. But if reason is solid at a low level, but not at a high level, what is the boundary of its reliability? If there are no indivisible points in space, how can there be any sharp edges between the level of reason that is permissible

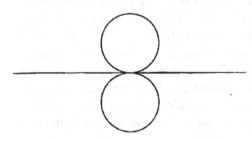

and the depth that is impermissible? Under their premise, the whole thing becomes a smudge in which theory and experiment become confused, and fudging is the way taken.

Let us consider some important aspects of the spacial infinitesimal. (In Chapter XXIII, the temporal infinitesimal will be considered).

This infinitesimal takes up space. If it did not, another line could be inserted between a geometrical tangent line and the curve that it is touching.

Sometimes, the positioning of the points can make a big difference, even though it is not apparent on the basis of what we can perceive or our instruments can detect. The drawing above is ambiguous.[105] If it is a complicated figure, analysis is quite easy to interpret, for it is easily resolved into two circles and a line touching both, i.e., tangent to each of them. But the analysis is more difficult if it is a single figure, a line with two loops on opposite sides.[106] Both the complicated and the simple figure look alike, and yet they are very different. The two circles in the complicated figure only rest on the line, but this difference could not be discernable with the most accurate electron microscope. It is at the level of the infinitesimal that the difference exists. This might seem trivial, but, as we will see later on, it is very important whether or not a line merely touches.

In short, two different drawings can look alike with the distinction lying in the placement of an invisible point.

The length of a finite interval is determined by its two end points. If these end points were themselves intervals, then there would be no stability to the boundaries of a given interval. Its terminal points would be like disappearing mirror images—only without the loss of resolution.

The dominant idea of the last century was that this was the same with respect to any two points. This is well expressed by Professor Huntington in his study of the continuum: *"If a and b are elements of the class K, and a < b, then there is at least one element x in K such that a < x and x < b.* Any series which has this property is said to be *dense*. Between every two elements of a dense series there will be at least one and therefore an infinity of other elements; so that no element has a successor, and no element a predecessor."[107]

In contradiction to this notion, there is necessarily a first element in the number line, the groundwork for any continuous series. There are neighboring points.

If α and α' are two neighboring infinitesimals, then their difference, $\alpha - \alpha'$ must be an unqualified zero. There is no position that could be possibly occupied by this difference of zero. It might be possible to speak of a group of five neighboring infinitesimals and concentrate attention on three, leaving two out of consideration. One might even attend to all of them, leaving none unattended. But this "none" would not be like the 0 mark in a number line.

In absolute space, there is no place for an 0 which is not a position. Zero can be the result of an arbitrary decision to designate a certain infinitesimal as "0"; zero can also be used to represent the deprivation of some thing, some action, or some attribute; for instance, absolute zero in temperature is usually taken to mean the complete absence of any molecular motion. A point in empty space is still a position, whether or not anyone has located it in some system of coordinates.

But the simple subtraction of 5 infinitesimals from 5 infinitesimals does have meaning. It means that there is nothing left to account for. It can mean that one must stop considering that collection of five. It cannot mean their extinction or transformation into something else. There exists both an absolute zero and a relative zero. This distinction is very important for us, for we shall discover later on in this chapter that subtraction and division are not completely synchronous, i.e. that division is not simply a short method of making several subtractions in every case.

There cannot be any ratio between any two spacial infinitesimals with any other value than "1". But this one cannot be the "1" of unity which may be resolved into rational and irrational intervals. This can only be the "1" of identity, which cannot be divided into fractions representing less than the number of infinitesimals under consideration, for there is nothing smaller than an infinitesimal. If one were considering a collection of 12 infinitesimals, it would be permissible to speak of 3/4 of that number, which would be 9, but one could not speak of 3/8 of that number, for there is no such thing as 4.5 infinitesimals.

It is possible to label one infinitesimal as zero and a finite number on one side of it as positive and on the other side as negative. But, as stated before, there is no zero in absolute space. (For reasons which will not be clear until later, subtractions like +3 -(-6) = +9 are not always valid).

Therefore, the use of finite numbers in connection with infinitesimals must be done with great dexterity.

As a practical matter, the fact known since before the time of Christ that rational quantities do not suffice to fill in a unit, but irrational quantities are required, means that there is something which is part of the unit but does not fit into it in a manner fully intelligible on the level of the finite. Reflection on this fact naturally leads to the idea of the infinitesimal. But since it concerns the imperceptible, error is easy. A case in point: proponents of the standard idea sometimes use the term, "infinitesimal." But their meaning is very different from that in use in this book. Edwin Bidwell Wilson was a prominent mathematician early in the last century. It was he who wrote the first textbook on the modern system of vector analysis.[108] In his work, *Advanced Calculus*, he wrote that "an infinitesimal is a variable which is ultimately to approach the limit zero . . . ."[109]

On the contrary, the actual infinitesimal does not become zero, but maintains its identity. The spacial infinitesimal is a location. This infinitesimal is beyond number. Take any interval. This interval is finite. Given signed numbers, once a 0 point has been picked, then an interval can be measured. The infinitesimal at the end of the interval opposite to zero has a number also, for instance, 2. But that number, that evaluation, exists only because its place in that extent is determined by the placement of the zero. Indeed, that same point could itself be chosen as position 0, with the point presently marked by "0" being labeled as -2. In short, the necessity lies in that which must exist after the point has been asserted.

Since the "infinitesimal" as defined by Dr. Wilson tended toward (and even became zero), there were various grades among his infinitesimals. "If any infinitesimal α is chosen as the *primary infinitesimal*, a second infinitesimal β is said to be *of the same order* as α if the limit of the quotient β/α exists and is not zero when α [approaches] 0; whereas if the quotient β/α becomes zero, β is said to be of *higher order* than α, but of *lower order* if the quotient becomes infinite."[110]

But there is only one kind of spacial infinitesimal, the one that is no more than a position or location. For the same reason, theorems, like Duhamel's, are ruled out at once. According to that theorem, should the sum $\Sigma a_i = a_1 + a_2 + \ldots + a_n$ approach a limit as n becomes infinite, the sum $\Sigma b_i = b_1 + b_2 \ldots + b_n$ with each b_i differing uniformly from the correspondent a_i by a higher order infinitesimal, would approach the same limit.[111] Contrary to Duhamel's theorem, there is only one grade of infinitesimal.

Back in the 1960s, Abraham Robinson came up with a new type of number, which he called a "hyper-real number." This was defined by him as a real number plus an infinitesimal. [112] His idea of the infinitesimal was incorrect, as he considered it to be a number on its own and therefore with the same metaphysical status as a real number. He wrote that just plain zero qualifies as an infinitesimal. The truth is that real numbers are finite intervals on a number line that is ultimately composed of infinitesimals. What he ended up with were certain kinds of irrational numbers. And, since real numbers include the irrationals, his hyper-real number means that a certain kind of real number possesses a greater magnitude than does the comparable normal. This is a misnomer, for such numbers can be smaller than the normal, also. (His basic discovery was sound, however. It was found independently by the author. Abraham Robinson's precedence is hereby acknowledged. The correct form is presented on page 83 under the name, "positional magnitude." Robinson was inspired by Leibniz's idea of infinitesimals, which, it will be shown, provided a poor model.)[113]

Having examined some important errors, let us now concentrate attention on the linear interval. Just for the sake of the argument, suppose that there were a minimum linear interval. This would mean that there could exist an interval which would not be capable of further division. Then, two of these minimal finite intervals could be added together. One end of the stack could be designated as "0" and the other end as "1;" this could make the place at which they join together equal to "1/2". But since both are minimum finite linear intervals, there could be no "1/4." Similarly, if three of them were stacked together, the two places at which they were joined would be 1/3 and 2/3, respectively, but there could not be any such thing as ½ for such a stack, since the second minimum interval could not be capable of division. If five were stacked together, there would be one fifth, two fifths, three fifths, but no one half or one third. Furthermore, the parts would only be rational. There would be no irrational intervals, since the number of intervals, however great, would serve as a denominator for every constituent. That would be the case as long as the stack were finite. And finally, only if an immeasurable quantity of these minimum finite intervals were stacked upon one another could there be the divisions of the common interval. But such a stack of any finitude, however small, would be of fantastic length, not a finite interval of definite extent.

Any finite unit must possess within it the possibility of generating irrational numbers. If the minimal interval were possible, this would mean that there would be no rational divisions; that it could be broken, but not divided. It would mean that successive minimal intervals could come up with $\sqrt{2}$, but not 1.4.But, in order to reach this former, it must first pass the latter; this is shown by the approximate, 1.41421356. Consider an irrational magnitude within the bounds of a single minimal unit; $\sqrt{.2}$. In order for that to exist, the point .44721 would first have to be reached. The minimal interval would have to be the same as a point, which is contrary to reason. Even a string of infinitesimals is capable of rational division. It seems that a minimal interval or length cannot be found by just eliminating rational divisions. (This subject will be resumed towards the end of this book).

Now, let us turn to the curvilinear. Just for the sake of the argument, suppose that there were minimal curvilinear intervals. This would mean that there could exist an interval which would not be capable of further division. Then, two of these minimal curvilinear intervals could be added together. One end of the conjoined pair would be the beginning and the other end could be designated as "1"; this would make the place at which they were joined together equal to "1/2." But since both are minimum finite curvilinear intervals, there could be no "1/4." Similarly, if three of them were combined, the two places at which they were joined would be 1/3 and 2/3, respectively, but there could not be any such thing as ½. For such a connectivity, the second minimum curvilinear interval could not be capable of division. If five were linked, there would be one fifth, two fifths, three fifths, but no one half or one third. Furthermore, the parts would only be rational. That would be the case as long as the connectivity were finite. And finally, the ultimate constituents of the minimum curvilinear unit would have to be irrational magnitudes, since there could no more further rational division. But, this would not be possible, since the irrational magnitudes include within them rational magnitudes. As with the straight line, the question as to whether there is a minimum curvilinear interval cannot be answered that way. (This discussion will be resumed later, also).

The straight line cannot be defined in terms of curved lines, nor can curved lines be defined in terms of straight lines. They are very different. There is no way to define either kind in a way that someone who did not already know what they were could understand. Euclid's definition of the straight line in terms of the points lying evenly on one side meant that it had a single, consistent direction. In those terms, one could refer to a curve as a line which changes direction regularly at the level of the infinitesimal. But, as implied above, the concept of direction is unintelligible in the absence of that of a line; so, although one might be able to distinguish straight from curved lines in that manner, or even straight from bent, there is no way that one could define a line to someone who had never experienced one.

Those familiar with mathematical literature frequently encounter certain figures of lines bending inward upon themselves, filling all space and terminating on a point. Peano, Hilbert, and others have imagined those figures.[114]

In order to fill any space at all, these lines bending within each other must first start to touch one another. Once this takes place, that and the successive intervals involved no longer have a finite difference between them. Since they have already broken below the level of the finite interval, it is no wonder that they can fill space and terminate on an infinitesimal.

NO AREA FILLED

To understand this fact more graphically, imagine a square figure like the upper one drawn on the left. Let their lengths be one English foot. Now let us begin to draw lines parallel to the base, with the distance of each succeeding line from the base being one half greater than its predecessor. Theoretically, if one were to mark off the first 1/2' above the base line; the second, 1/4' above its predecessor; the third 1/8' then 1/16', then 1/32', 1/64', 1/128', 1/512', 1/1024', etc., one would never be able to reach the 1' mark. Since these lines are hypothesized to have no thickness, it is theoretically impossible for them to fill space.

TOTAL AREA FILLED

Now, look at the lower figure on the left with the same base and sides. Lay a line segment of the same finite length as the base upon the latter, i.e, next to it. Then lay successive lines upon each other. Theoretically, the square would be completed and filled. Although no line had any finite breadth, every point would be filled.

Why is it that in the second case, the space would be filled but not in the first?

In the first case, one proceeds by division; each successive line cuts the remaining space in half. Such a process appears to be endless. Spaces are enclosed, but never filled. It is a potential infinity. In the second case where each successive line is placed on top of its predecessor, no space is left between them. Each time a line segment is added, the remaining distance on the vertical lines at the side gets infinitesimally smaller. This is subtraction. Division, it will be recalled, is an operation in which the factor of a dividend is found when given another factor, the divisor. There is no zero absolute with division. It must stay within the number line, never on its border. Subtraction, on the other hand, can totally exhaust all quantities. It can complete the square. It can fill every space.

This is a remarkable conclusion. Although every division in the first case remains completely within the realm of the finite, division can only jump spaces. Because it seems that space must always remain, the process appears to be unfinishable. The other one can only proceed infinitesimal by infinitesimal. It is unmistakably finishable. Putting it differently, through the process of division, there cannot be a minimum interval; through subtraction, all intervals can be eliminated. In a later chapter, this will be shown to be the answer to some paradoxes of Zeno.

This introduces such questions as, What is actual infinity? Is the quantity of infinitesimals within a unit endless? If not, then there must be an end to division, even if the process does not indicate it. Since Weierstrass and his friends apparently considered division within the unit to be inexhaustible, we wonder whether that might be the root of the inadequacies of their theories. The answers will begin in the next chapter.

CHAPTER VIII

WHAT IS ACTUAL INFINITY?

PART 1—THE IMMEASURABLE

The dominant mathematics of the last century and a half has implied that the points in a single unit are infinite in the sense of being endless. Such loose logic has led to the hypothesis that the part is equal to the whole.

Compare a line segment three inches long and another one inch long. The greater length has as many points as the shorter one in its first inch alone. The larger is greater than the shorter, even though the points within the shorter one are innumerable. When placed against each other, the one inch measure is matched point by point by the other measure three times over. It would be worse than Sisyphean to attempt to show this on a one to one basis. But they can be compared through the simultaneous presentation of the two lengths. The three-inch segment comprises a greater infinity than one of those units considered by itself. Would it be technically correct to call it three infinities? If it is, then Cantor was correct in speaking of one infinity being greater than another, though not in the way he believed it to be.

Cantor believed that his infinities existed eternally in the mind of God and were therefore numbered.[115] With this premise, he did not need the distinction between numbers and quantity. There would be no need to talk about any potential infinity either, since they would be already there. That way, he could speak intelligently about sets with infinite members. This belief of his that God had shown him the way gave his ideas a certain concreteness.

The typical 20th century mathematician accepted his conclusions while denying his major premise. They regarded the number line as a mere figure of speech into which endless "numbers" can be assigned. This is incoherent. Since it puts points outside of space, there is no way that their definition of infinity as having a part equal to the whole can make sense. "Part" is a spacial concept.

Not only have they disposed of the ladder that got them there, they risk forgetting that they ever needed it.

Recall the square composed of equal straight lines segments stacked one upon another, horizontally, until there were as many lines as there were infinitesimals on the base line? Let each of these lines be placed end on end. This would be an infinity of finite units. No human enumeration could reach it.

A rectangle made of such lines with the shorter side the same length as the side of the square in the previous paragraph would contain a "greater infinity," since the longer side could have any length we desired.

To understand actual infinity better, let us return to definition. Two characteristics of actual infinity have been noted: the first is that it be immeasurable, beyond man's power of enumeration. The second characterization of is that it be beyond not only numbers, but that it be endless. By this characterization, neither the complete unit nor any compound, however staggering to the imagination, qualifies. The equilibration of immeasurable, but surpassable quantities with absolute endlessness, therefore, is false.

An infinitesimal is indivisible; it has no parts, no area. Cantor denied the spacial infinitesimal. He thought that its quantity within a unit would be so great that it could never add up to a unit, in fact could not be contained even by one of his transfinite numbers.[116] But, as we have seen, the infinitesimal must occupy space; two different locations cannot coincide; they can, at most, be neighbors. The infinitesimal is the smallest element of space that exists; it is an indivisible location or position in space.

The quantity of infinitesimals in a unit is immeasurable, but not beyond all bounds. It is definite. It is infinite only in respect to its being immeasurable and outside the realm of the finite. Since every finite length is composed of innumerable positions, every part of its extent along a unit is taken up.

Cantor wrote that "non-zero numbers ς (in short, numbers which may be thought of as bounded, continuous lengths of a straight line) which would be smaller than any arbitrarily small finite numbers do not exist, that is, they contradict the concept of linear numbers."[117] Cantor was right in thinking that if there were an infinitesimal, it could not be linear. In fact, it does exist, but below the level of the line. The unit's length, it will be recalled, was distinguished from a mere assemblage of infinitesimals by possessing the potential of irrational magnitudes.

As stated above, Cantor implied that the points in a unit could not be unlimited. This is an important admission from him, the truth of which is undeniable. Consider the square in the last chapter in which each line within it was placed half the distance remaining after the one before it had been positioned: it was shown that according to the idea of infinite division, it could never be completed. It is now clear that this process of cutting in half the remaining distance would grind to a halt. Eventually, the immense quantity of infinitesimals along the side would be run through (and passed). What was evident with subtraction would also be the case with division. It is just that we humans cannot calculate it, since we do not know how many infinitesimals exist along the side of any unit. The process of division does not stand on its own. Without the understanding provided by knowledge of the infinitesimal, it is paradoxical.

The denial that .999 . . . 9 could ever equal 1.0 does not change. Even if one were able to do the impossible and decipher all the positions between 0 and 1 in a unit, the sequence $9/10 + 9/100 + 9/1000 + \ldots 9/10^n$ would still not be able to reach 1, exactly. The terms for the last two positions would have to be $9/10^{n-1}$ and $1/10^{n-1}$.

It is also the case that a Being who lives forever would see the asymptote and curve meet; this is because there are only so many infinitesimals for the curve to cross until it joins with the line. In fact, this realization affects potential infinity itself. But that will be postponed until later.

It follows, therefore, that there are two kinds of actual infinity; the type that is not-finite and the type that is without end. Both are defined negatively, because, their existence is inferred from that which is ultimately derived from the senses.

Let us call the first type of infinity, "immeasurable." Its symbol is "$\mathfrak{I}$", which is Old English for "I." Most computers have it.

Consider the optimal unit size with its complement of points beyond measure. We will not define what is meant as "optimal" until much later in the exposition, since it is better that the basic ideas behind this type of infinity be understood first. Let us designate the quantity of points within that unit as "$\mathfrak{U}$." This is Old English for "U," a form of which can be found in most computers.

Next, let us define the uncountable extent which would exist when a quantity of units equals the quantity of infinitesimals within a linear unit as "$\mathfrak{I}_{\mathfrak{U}}$." This symbol is also written in Old English. It is considerably less vague than characterizing it as being greater than any number of which we can conceive. (Of course, $\mathfrak{U}$ is also immeasurable, but this symbol is used to stress how stupendous an amount it is, how unfathomable the attempt to plumb it through visual imagination).

Can this be used as a method for calculating higher and higher infinities? Let us see. The line extended from the unit square is an infinity. The extended line made of as many lengths of 1-unit radii as there are points in the perimeter of its circle would be $\pi \mathfrak{I}_{\mathfrak{U}}$. The extension of a rectangle with a longer side equal to $\mathfrak{I}_{\mathfrak{U}}$ and its shorter side equal to $\mathfrak{U}$ would be $\mathfrak{I}_{\mathfrak{U}}(\mathfrak{U})$. Symbolical calculations can also be made for full objects, like the cube, the cone, the cylinder, and the sphere.

This type of actual infinity, the countless and immeasurable, stands in opposition to potential infinity, as does the other actuality, which will be discussed in the next chapter. The symbol of potential infinity is to remain "∞". It consists of the attempt to count what finite beings are unable to finish. It refers to an action and not an abstract quantity. Cantor implicitly thought he could exclude it by supposing that all numbers have already been counted by a Divine Being. But potential infinity is a true reflection of the human condition, and should be acknowledged. The best example of this is the natural number system itself, whereby given any number, N, a greater number, (N +1), is possible. With that concept, it is possible to understand that magnitudes greater even than $\aleph_{\aleph}(\mathfrak{U})$ are possible. Although never actual, potential infinity has the potency its name suggests. No line can be anything more than a potential infinity. Contrary to Cantor, one cannot assign what one does not have.

To repeat: Immeasurable infinity results from man's incapacity, not merely from an inherently unfinishable task. Potential infinity can be applied to either.

Because of this immeasurable infinity, there are magnitudes within the unit(s) which cannot be located by finite means, even in principle. These are not root functions of rational numbers like $\sqrt{2}$; nor are they directly derived from geometrical figures like π; nor are they the culmination of the study of the exponential function like e. Consider a square that is one unit in length on each side. Now, imagine another square inscribed within it that touches it throughout, leaving no space between it and the outer square. The difference in the length of each side would be two infinitesimals. The ratio between this length and the unit length is not finite. It is not commensurate with a finite fraction of the unit. It is irrational.[118]

Its nearest approximate is exactly what it clearly is not: 1 unit. Since it cannot be distinguished in finite terms from the unit, it is not identifiable in those terms. There are vast quantities of points within the unit like this, each of them the potential endpoint of a magnitude. Let those which cannot be numbered either directly or through approximation by finite procedures be known as *positional magnitudes*. Like the positional points defined in Chapter IV, they cannot be numbered.

Now, an unexpected consequence of immeasurable infinity will be considered. Just as the limitations of space rule out a unit with unlimited points, so they prevent the possibility of certain figures which superficially appear to be plausible. On the left, steps are represented as running along the diagonal of a

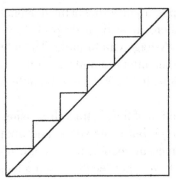

square with the top level flush with part of the upper horizontal side of the square and the lower rise flush with part of the left-vertical, six steps altogether. The total length of the rises and levels would be double that of either side of the square. Since the length of either side is 1, their sum has to be 2. At a 45° angle, half of the points come from the vertical; half from the horizontal. The total length of the steps is double the contribution of one of the sides of the square, i.e, two.

But we know that the true length of the straight diagonal is the square root of two. This would be the case, regardless of how small we make the rises and levels. Admirers of the limit approach might suspect that the smaller and more numerous the steps, the closer the sum would get to √2. But this is not the case. No matter how tiny the rise, the sum of their descent is 1. The same with the levels; the more insignificant their length, their sum is also 1. Even if, one could descend to the level of infinitesimal steps, their sum would remain a stubborn two—providing that enough points were used to form the steps. The length of the diagonal, therefore, cannot be analyzed into contributions from the horizontal and the vertical steps. Any attempt to call the diagonal an "infinite stair-step" is laughable. Any intellectual trick which departs from a single straight line must fail.

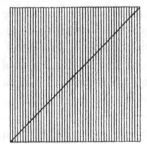
Consider the diagonal and square represented on the left. The vertical lines are all one infinitesimal in "cross-section" and are all placed one next to the other so as to cover all the space. That being premised, it follows that every point in the base and top is covered; that there is no blank space in the interior of the square. From this it follows that every line goes through the diagonal. Yet, the diagonal is longer than any of the sides.

How is it possible for every point on the diagonal to have a line through it when it has more points than the base? Would the analyst be forced to consider infinitesimals of more than one size, the larger one a multiple of the smaller one?

As it turns out, this is not necessary. Obviously, the answer has to do with locations of the points along the diagonal. To understand the answer better, let us consider a similar problem which does not involve irrational lengths. Take the rectangle with a base of 3 and a height of 4. Its diagonal is 5. With this diagonal, there are two kinds of changes: the left-to-right variety and the up-to-down variety. Infinitesimals are just positions. *Obviously, there are more changes of position when there is both an up-down and an over-across action than there is when there is only one of these.* Starting from the upper right corner of the diagonal, this means that for every level change of 3, there must be a drop of 4. We discover how much longer it is by applying the Pythagorean formula. The total length of the diagonal is 1 unit greater than the height and 2 units greater than the base. If, *impossible*, we could perceive infinitesimals, we might be able to count them, instead.

But the diagonal's length is not simply a function of the abstract dimensions of two of its opposite sides (3 and 4, respectively), but of its shape and area. These dimensions are enough for purposes of measurement, but they are not what it is all about. Two lengths 3 units long and two other lengths 4 units long standing right next to each other would enclose no finite diagonal area. Since the individual infinitesimals themselves do not have any shape, we cannot picture it very well—only calculate it.

The area of the rectangle is constituted by its positions. *They are all there.* Room, therefore, exists for the long diagonal. And, lest we forget: a five-unit line can originate outside of a rectangle or a triangle. How strange the ordinary!

Returning now to the diagonal of the square, which is also the historic basis for the discovery of irrational numbers, we can see how important its area is, for the calculation depends upon the square root of the sum of the *squares*. As was shown earlier, areas are different from lines. Also, keep in mind that infinitesimals are not arrayed in space; they are that in which arrays are arranged.

In the next chapter, the type of infinity which is independent of the limitations of human nature will be introduced.

CHAPTER IX

WHAT IS ACTUAL INFINITY?

PART 2—ENDLESSNESS

In the last chapter, the infinity which exists relative to our capacities and the machines we make to assist perception was discussed. Is there an infinity that is absolute, beyond not only numbers, but with no termination whatsoever?

The symbol for endlessness is "𝔈". This is the Old English for "E"; the form of it on the reader's computer will do. It stands for endlessness. Interestingly enough, Cantor says that the limitless does indeed exist, even beyond his hypothecated transfinite numbers.[119]

If something is truly endless, there cannot by definition be the fabled point at infinity. This is so by necessity.

Actually, there is more than one kind of 𝔈. But the one that is to be discussed in this chapter is Space.

Actual infinity in that sense was discovered from the idea of space. Percepts indirectly disclose the existence of space. Two or more objects cannot exist at the same place and the same time; therefore the display of existence as manifested in the form of percepts does not show them all jumbled together in the same place, but as occupying distinct positions. Stereoscopic vision reveals 3-D directly. Even if a man were born with only one eye, he would still perceive depth because whenever he moved his head he would experience parallax. A man born blind could discern distance through the differing sounds; for instance, in his infant state, what the increasing loudness of his mother's footsteps meant, as she approached him. Even a person born deaf and blind, like Helen Keller, would not be at a loss as far as the recognition of space is concerned. An infant's arms and legs move, as a result of which there is a consciousness of room; it will also show that the human body and the crib are more solid than the air, the latter of which may even seem to the very young as made up of absolutely nothing. The blind man who had reflected about the breeze could realize that air is not empty but only less solid than the ground, which in turn is less solid than the stone wall at the property's edge. It is from such experiences that one can derive the idea of density and empty space. Distance can be discerned through any localized sensation; a muscular contraction in the leg would be enough to point out its separation from other places in the body. These realizations come from perception of the outer world and do not require any explanation in terms of an innate idea, although the faculties of perception must be formed in advance in order to receive impressions of sense. For instance, the sensation caused by the pressure of the air against an infant's leg as it moved about would indicate, not only room, but distance. There, the ability to sense something distant is built-in, innate as it were.

Space is simply a fact which the perceptual mechanism takes into account; this it does, whether one senses oneself stepping on a stone, feeling warmth while drinking a heated liquid, or looking down at some rolling hills from an airplane. An animal also takes such things into account. A dog knows what it is to be in a cramped area, or to have plenty of room to run about in. But, as far as this writer knows, they have not formulated the idea of extension itself.

The notion of space without any objects in it can result from remembering something different in a scene, wherein we understand that what was once present in some place is not there now; by thinking about this, we come to understand that the distances in all directions are cognitively independent of what we perceive; that the house we see over there could be removed, but its present distance from us and the space it presently takes up would still be there, that it is the same for us and the land itself on which we are standing, and so on. The realization of this is not itself innate, but is an inevitable consequence of thinking about distance. The induction and subsequent acts of deduction are easy. (Later on, in Chapter XXIII, it will be shown that the very idea of time does not derive from perception, but is innate from its beginning in our consciousness).

Put differently: we do not directly experience space, because it is imperceptible. We are conscious of differences, and in the homogeneity of space, there is no perception of difference. It cannot be detected with the most accurate instruments, since these are extensions of the senses—built, operated, and interpreted through their aid.[120] Internal sensations give us knowledge of distance; and from that, the idea of space is not far away. We also infer its existence from what we perceive of the objects in motion. This inference involves both induction and deduction together. From the characteristics of objects standing and in motion, we draw the conclusion that they themselves do not fully explain what we are sensing. From this lack, we generate the idea of room. From that, through further abstractions we discover such properties as the homogeneity of space, etc.

We become aware of the infinitesimal of space through contemplation upon the meaning of continuity. The fact that it is for us an abstraction does not make it any the less factual.

The infinitesimals of space follow from each other without a break. Neither do they have any shape. They are locations, but not areas, for the latter imply more than one point. Each location is different, but they cannot exist in separation from one another. Nor do they interact.

Yet, the mind can abstract this idea, though the infinitesimal can never be perceived, even with the greatest magnification. Expressing continuity through the language of the opposing idea of the discrete allows one to focus upon what would be inexpressible, otherwise. They make the geometrical line possible. Right after stating his first definition in Book I that a point is that which has no part, Euclid defines a line as breadthless length.

Once having grasped the concept of space, the steps are short to the idea of an actual infinity in the fullest sense of the word. Children often reel when they come to that thought, for they know it is too much for them. They know that they will never be big enough to handle that. Some withdraw from it, turning instead to the companionship of their fellows. Others deny the validity of what they have thought, supposing that the only thing that is real is what they can perceive and maybe a little beyond that. As grown ups, they will talk about relative this or relative that, or of "in the context of."

Then there are those who would like to know more. Thinking men reasoned from the fact of density that there must be empty space; that differing materials which occupy the same volume must have less matter in them and, therefore, more of the available space must be empty. Others reasoned similarly from the diminishing of radiant energy as it spreads outward from a source—and so on.

Euclid's idea was that space is a potential infinity. But that is not enough to characterize it. ∞ is restricted to mathematical approaches to specific values or even to indefinite increases. But space itself is endless. It consists of locations beyond all thought of enumeration. Furthermore, each location is an infinitesimal, and they exist, regardless of whether men are capable of eventually positioning them in some sort of a coordination system or not. It is endless locations going in all directions, the part that is empty and also the part that has matter and energy in it, for they, too, must take up space.

Stated differently: each location does take up space. If it did not there would be more than one location at the same place, a manifest contradiction. Together, but mutually excluding each other, they are everywhere in both senses of that phrase.

Some might think that this is no more than a bunch of zeros going off in all directions, the sum of which is nothing. But each one of these supposed nullities is a unique location, different from all the rest. They are everywhere; but with respect to being unique and found nowhere else, each one is unrepeatable.

These unique locations exist always and everywhere; therefore, they cannot be conceived as something merely relative to human making or doing; to the contrary, far from being locked within the human context, they circumscribe the human condition. It would take a poet to characterize the largeness of this idea.

In his well-written book, *The Ten Assumptions of Science*, Glenn Borchardt has objected that "'an element of spacial extent' can represent one of two possibilities: either it is something or it is nothing. If it is nothing, then it is indeed empty space and hardly could form a connector of any sort. If an 'element of spatial extent' is something, then it must have matter within it and therefore it can be considered to be an object."[121]

As we have seen, the choices articulated by Dr. Borchardt are not exhaustive. Although space is not the least bit material, it still exists. Each of the infinities of infinities of infinities of spacial points constitutes a distinct location. Although there is no connector linking any two neighboring infinitesimals, the connection between two points which are not next to each other are the points which lie in between. A material glue is not required.

This is an absolute; as such, it makes relative spacial determinations intellectually cognizable. It is because none of these infinitesimals can waver that men may symbolically construct coordinate systems, whereby one of these locations stands as the reference to which others are related. The drawings made by men are not exact; the planet in which they are situated is in motion. Nonetheless, the knowledge that there is this ultimate fact allows men to make these constructions without great presumption. If everything, including the ultimate locations, were in flux, such efforts would make the outcomes of the gambling table seem like certainty by comparison.

Recently, the writer, Russel Moe, has asked, "Tell me again, how do we objectively see beyond travelled distances and recorded history?"[122] This adventurous questioner might be asked: How do you know you can rely on "recorded history" when you weren't there to observe it? More to the point, why is it that the reports of people who say that they have been to places you have not can still be believed? To this, the answer is usually given that a small amount of extension beyond actual experience is permissible.

A limited extension? How far is that? If a person enters a forest the first time, how does he know that the ground he steps on will not fly into the air, carrying him with it, then dump him on the ground, which in turn rises up, etc. ? Is it only because the ground has thus far acted in the familiar fashion? Was Hume right in thinking that causality is only a habit? But Hume contradicted himself, for a habit would be a cause, would it not? And causality is a grand principle which extends beyond known experience. Does Mr. Moe believe every "experiment" reported by scientists? The law of the lever still works in the manner described by Archimedes who lived before the time of Christ. How do we know that law will continue to operate on the levers we have not yet tried?

The answer is that is that a reason may be extended until and unless it is confronted with plausible objections. These, it must answer before it can go forward. One cannot simply state arbitrarily that there is a barrier to the extension of space. In the absence of knowing that it in fact exists, such a claim would be cognitively worthless. Neither can one say that a barrier is possible, unless one can show on what grounds it is possible. There have to be some facts in support of the barrier before it can be considered a possibility. For an extended discussion of this issue, see the author's book, *The Stance Of Atlas: An Examination Of The Philosophy Of Ayn Rand.*

What plausible objection can be made? That space is an 𝕰 cannot be denied. For to deny it is to mean that there is some location beyond it, and that would be a contradiction.

Bernhard Riemann started the talk about space being finite but unbounded.[123] According to that theory, the universe is like the surface of a sphere; the latter can be said to be unbounded, because there is no force to stop a person from traversing from one part of it to another. But even an intelligent creature the size of an ant crawling on the outside of a spheroid would recognize from its own thickness that space extended farther than the bottom of its feet.

In order to prevent this absurdity, the intelligent being in a finite but unbounded universe would have to be inside, rather than outside, experiencing a space that is fundamentally concave. But actually, if a person had to spend his life within a steel globe, he would experience confinement, even if he had magnetic shoes which allowed him to walk all around it. Imagine that the spheroid started to expand. Would not a person begin to experience a flatness in his path, evidence of expanded horizons? Suppose it started shrinking and his paths began to curve more, causing him to feel more bound. Even if it did not change at all, the perspicuous investigator would be bound to ask what was beyond the closed loops he was traversing.

Why did Riemann conceive such a weird model? The reason is that he was convinced that there could be no-action-at-a-distance, that physical attraction must be a form of contact. His solution was to return to the old Aristotelean notion that the universe had to be a plenum, that it was full, which required that the universe be finite.[124] How could that be even thinkable ? He conceived it with the twist that there are no straight lines, that it is like the surface of a sphere as found in what was then called "solid geometry." But Reimann's idea is incoherent. One of its incoherencies is that, unlike a genuine sphere, his could not have a diameter, for that would mean that *there was something straight.*[125]

More than seventy years ago, J. J. Callahan, President of Duquesne University in a work on the famous parallel postulate dispute, wrote these words: "It becomes really amusing to see Riemann's follower, Einstein, later explaining this. He illustrates it in this wise. If we take a number of discs and spread them out on the surface of a sphere, the number that will fit on the sphere will be limited, and the sphere is thus shown to be finite. But if we take only one disc we can push it round and round and never come to a stop. In this sense the sphere is unbounded. Can it be that anyone would ever fall for such rubbish? One might thus fool an animal. A tiger might walk round and round in a cage apparently never seeking an exit, but could any human being ever run round the same path and persuade himself it was unbounded?—not outside of a *maison de santé.*"[126]

Finite space is bounded, regardless of the prisoner's state of awareness. To prevent such an absurdity from being apparent, it is supposed that the closed universe is so large in proportion to the size of men that they, with their little longevities, will never be able to feel circumscribed.

But even if, *impossible*, this were so, the logical question is: What lies beyond this confinement? It is impossible that there would be any bound to space, however, for if it were, it would have to take up space, which would in turn bring up the original question: What lies beyond that? This truth cannot be evaded by supposing that there are no straight lines, only closed curves. Unless one were to put oneself under hypnosis, one could not honestly deny more curves outside the boundary. And the more curves, the less plausible the excuse for a plenum. And then comes the question, what is the distance between these outer curves? which brings up the question, then why can't there be a straight line across this distance equal to it—a diameter?

Furthermore, how would the person under the spell explain how it is that others disagree with him or her? Suppose this person tried to answer that since the universe is so large, the curve would appear to flatten out, giving the illusion of straightness. But that would be an admission that this person knew what such a line was? If it cannot exist, how is it that men can arrive at it so easily? Straight lines are not repugnant to the mind like square circles. It is only with reference to straight lines that we can recognize a curved line.[127] Suppose they were to hypothecate an asymptotic process, but one in which the straight line to which it tended could not exist, a goal which is in reality a contradiction. But if the approach to this were endless, how could it be kept within the bounds which the Riemannian requires? If almost straight lines exist, wouldn't they go out quite a ways before they started to curve back? And since the line tends infinitely toward the straight, are they not left with at least a potential infinity? And if that, they are at least back with Euclid. Furthermore, if the line can never be straight, they are also back to endlessness. Why? Because if the points to be crossed by the line in order to reach the asymptote are merely immeasurable, then eons from now they will cross, producing a straight line. And if they can never cross, then they have admitted the endless. Either way, they are defeated.

Encompassing all such bounds are the infinitesimals of space. They do not just occupy positions; they are these positions. Space does not have a shape. It is neither flat, nor round, nor hyperbolic. Completely homogeneous, it is that in which every shape that can be formed must exist. It is everywhere.

CHAPTER X

ANGLES AND CIRCLES

There is no mystery to the definition of an angle. Ray defines it as "the opening, or inclination, of two lines which meet at a point."[128] The two lines are called "sides" and the point from which they proceed is called the "vertex." The two sides must be finite; although they may be of any length, they must be of some length.

Any line meeting the vertex within the sides of an angle is called a "ray." The quantity of rays possible is dependent upon the size of the angle; the larger the angle set by the sides, the more rays that can possibly proceed from them.

The lines that intersect the vertex do not overlap, for that would mean that they are on more than one plane, an unevenness. Instead, they all share that point in common.

For the same reason, the quantity of rays stemming from a vertex is independent of the finite length of the sides. Whether the sides were both an inch in length or both a mile long makes no difference as far as the possible quantity of rays that may issue from it. Neither would it make any difference if one length were an inch and the other, a mile.

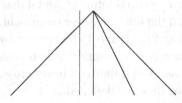

The quantity of rays within an angle of any sort is determined by the size of the angle established by the two finite sides.

It is impossible that any two rays coming out of the same vertex be parallel; for if they were, two vertices would be required, which contradicts the definition. Since they cannot be parallel, they cannot be stacked upon each other. They must splay outward as they proceed from the center. That is why the quantity of possible rays is dependent upon the size of the angle, a relational fact, and not upon the length of any interval set between equal lengths of the sides.

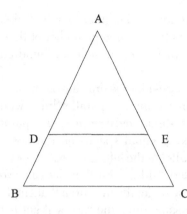

An angle of 0° is just a single line; so is a 360° angle. Why this is so follows because no two lines meeting at a single point can be stacked. That is an important way in which geometric lines differ from physical lines. The vertex of the latter are finite, often a ball-like structure. Knowledge of this fact requires that the definition of the angle be amended to read "an angle is the opening, or inclination, of two lines which meet at a point, or in the two cases of angles of 0° or 360°, or multiples thereof, a solitary line."

Consider triangle ABC represented on the left and above. Imagine lines stemming from the apex A and terminating at the base, BC. Their totality must pass through parallel segment DE. The maximal quantity of lines passing through BC can be no greater than that which must pass through DE. Owing to the DE being a shorter interval than BC, the quantity of rays must be less than enough to cover every point on the base. These straight lines, therefore, cannot spread over the entire space of the triangle. Furthermore, DE can be drawn closer to A. Regardless of how close DE is to the vertex, no fewer rays can go through it than can potentially touch the base.

Sometimes, Cantorians will claim that this means that there are as many points on the shorter line DE as on BC, that the two are on 1-1 correspondence.[129] This ignores the fact that rays emanating from A are splayed—which means that there must be more points on BC without lines going through them than on DE.

Question: Is there a minimum size to the triangle at the top? It might be argued that unless there were, the triangle could infinitely shrink until it became the size of an infinitesimal.

The answer is two fold: First, it could not shrink to the size of an infinitesimal because a triangle has corners and those are parts, which is contrary to the nature of that which must have none. In fact, not even three infinitesimals arranged with two neighboring ones at the bottom and the one on top sitting upon the other two is possible. The third infinitesimal cannot be positioned with one part of it on one of the lower ones and the other part of it on the other, simply because, once again, an infinitesimal has no parts. It cannot even be imagined to shrink that way. Second, and this is the decisive objection, for it holds regardless of how the triangle is set up, the maximum possible number of lines from A to BC is not determined by some fanciful smallest interval parallel to the base. What determines it is the angle at the top. The wider the angle at A, the greater the quantity of lines that can penetrate to the base. Given a certain angle, it does not matter how long the side of that triangle is.

All that is required is that the sides have some length, i.e., that they be finite. Given that fact, it is angle size which determines the maximum number of lines which can reach the base. The sides can be a fraction of an inch or a hundred miles long. It does not matter.

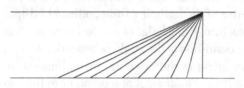

Consider the drawing on the left, representing two parallel lines with rays stemming from a single point making equal spaces on the lower parallel to the left. Let these lines be called "sides." Further, let the two sides be extended farther and farther to the left; also, let the rays from that single point be constructed one after another. At the same time, the rays will get not only longer, but also higher, reaching closer and closer to the upper parallel. The great astronomer, Johannes Kepler, believed that at infinity, a ray would finally be reached which would become parallel to the upper parallel. This, he thought, would be a point at infinity.[130]

But that would be impossible. In the first place, Kepler and the paradoxers who followed him assume that individual lines have no substance, that they take up no space. Only by assuming that can it be imagined that a line can be both parallel and incline with respect to another line. But the positions occupied by any straight line cannot be shared by another such line, except at one point. Since there is already an upper parallel line, all the rest must be spread out with respect to it. In general terms, there can be no coincident lines, for no distinct lines can occupy the same space at the same time. That is inherent in the nature of the infinitesimal, since each location must be unique. (Euclid did not use them.)

In the second place, parallel lines are lines which are equidistant from each other. If they ever met, they would no longer be parallel, but either meeting or intersecting.

In the third place, any of the diagonal rays would always be at an angle greater than zero from the point of origin. In order for the highest ray to be parallel, it would need to begin at some point below the point of origin of the diagonal rays, which would spoil Kepler's argument.

In the fourth place, the quantity of rays which could emanate from that point is definite. The maximum angle shown in the drawing is ninety degrees. No more rays could come from that angle than from any other right angle.

Please look once more at the drawing! If the parallels were extended on the right and diagonals were constructed mirroring the left, the new angle at the point of origin would be two right angles or π, or half a circle.

If the two parallel lines were removed, leaving the solitary point, and the maximum quantity of rays were allowed to intersect it, the result would resemble the familiar depiction of a sun burst. None of these lines would intersect and none would be parallel. The figure imagined by Kepler cannot be.

An angle is really a truncated sector of a circle. The arc has been removed, leaving the two sides joined at the vertex. That is why triangles are often drawn with little arcs on each corner of the figure.

Return to triangle ABC and horizontal level DE parallel to base BC. The higher up DE would be placed above BC, the more concentrated would be the rays joining the two levels to the apex. Yet, they did not and could not increase; the quantity that reached the base would be no greater than those at the apex. This means that not every point at the base was joined to one of these lines. The situation is no different with the complete angle that is the circle. The infinitesimals at the rim of the circle increase with the perimeter, but the quantity of rays is definite, regardless of the circle's size.

This is astonishing, albeit a consequence of the nature of an angle. This quantity is the same in any finite sector of a circle, however small. A one degree sector from a circle with a disk the size of the sun would send out the same quantity from its center to its periphery as would a one degree sector of a dime. Although the quantity of infinitesimals on the perimeter of the former would be unimaginably greater, the center of each would be a point, and a radius a millimeter around the center of the sun would contain no greater a quantity than that the same distance away from the center of the dime. The reason for this is that the rays issuing from the center are not parallel to one another; they cannot be stacked, and, therefore, must spread.

Two centers mean an ellipse. With the circle, the greater the length of the radius, the more the rays will spread—in fact, the spread is directly proportional to the square of that length. In this way, the perimeter's girth is exactly compensated by the radius' increase.

This also means that it is impossible for all the potential radii of a circle to be realized all at once. It is always true that a radius can be drawn from the center to any point on the perimeter, but owing to the stacking problem, not all of them can exist, simultaneously.

This is consistent with Proclus' definition: *although a straight line has the same measurement as the distance, the two are not the same;* the former is a construction, and the latter is a condition for its possibility.

Nor is that all. Since a radius can be drawn from any point, the selection as to which possibilities are to be picked for radii in any particular circle varies. It all depends upon which one is chosen first.

This shows once more the falseness of the idea that mathematics does not involve time. Here is a clear case of a temporal "before" and "after" appearing in the logical structure of mathematics.[131] The crucial element in time, the now, is accounted for by the impossibility of the simultaneous presentation of all the radii.

It also shows the geometrical significance of selection, a supposedly extraneous element, which their purists tell us does not belong to "pure" mathematics. This science does not subsist in some empyrean realm, totally in separation from conscious beings who have the power of choice. Its purity is of the same sort as any other human endeavor that is conducted without a breach.

What is true of the angle and the circle is also true of the point. There is a maximum quantity of lines that can be drawn through it, establishing a minimum angle. Only a certain quantity of lines may be drawn through it. This means that there must a minimum angle.

What is this minimum angle above zero? Suppose that the sides of a figure with one of these minimum angles were a million miles long. This would be an extremely narrow isosceles triangle. At the base, there would have to be quite a spread. Let the baseline be divided in two. A half mile up, construct another line parallel to the base. Let the midpoint of this line be joined to the midpoint of the base. A half a mile further up, another parallel could be constructed with a midpoint, which in turn could be joined to the line connecting the lower two midpoints. Let this procedure continue up the remaining 999,998.5 miles until the last half mile remains. Then continue as before with 1 feet increments until 2,639 of these levels have been completed. This leaves 1 foot remaining until the vertex. Then proceed with much tinier increments until just short of the vertex. It would seem that this line could be continued up to the top, splitting the final angle into two more angles. However, the premise is that the minimum angle has already been reached. It cannot be split.

What would happen if one attempted this? Let the angle be such that just below the vertex there is but a little space remaining open. A line could not be driven through it; it would have to split the unsplitable. Therefore, the smallest possible angle is an infinitesimal. This is the exception which proves the rule that any angle greater than zero can be divided. The exception does not contradict the rule, but amends it to hold only for finite angles. On the level below the finite, any angle greater than one infinitesimal can be divided. This exception is not a contradiction, but a condition. Furthermore, the fact that the minimal angle is the infinitesimal-width is a truth which could not even be approximated on the finite level. The minimal unit is not degrees, minutes, seconds, or some hyphenated subdivision, but the infinitesimal-width itself. Here, the term "width" is not taken to be a finite measurement, but the space taken up by an infinitesimal or location. This term was used, because of a poverty of language.

The angle between two of these lines can be as little as one infinitesimal, but their sides must be potentially finite and their lengths may be rational or irrational magnitudes. Trigonometric functions result from the sides and the vertex; as the world knows, these functions do not have to be rational.

Another relation between the circle and the angle which is of concern to the study of the infinitesimal is what many characterize as the "curvature of a circle."

This is defined as $\triangle\theta/\triangle s$, where $\triangle\theta$ is a tiny change in the size of the angle subtending an arc, and $\triangle s$ is a tiny change in the length of the arc s, as both get smaller and smaller. This ratio is equal to 1/r, where r is the radius of the circle. In other words, $\triangle\theta/\triangle s = 1/r$.

Starting with that, the standard argument is that (1) if radius r were allowed to increase to infinity, its curvature, 1/r, would decrease to zero; (2) that a circle with an infinite radius is a straight line, or to put it differently, a straight line is a curve with a constant curvature of zero.[132]

Neither is correct. Let us begin with the first consideration. Under the concept of infinity as the immeasurable, no radius, even one $\mathfrak{I}_{\mathfrak{A}}$ units long, could be the reciprocal of 0, *because the line can be extended beyond this*. If $\mathfrak{C}$ is employed, such a radius cannot exist, because it would have to be endless, which would render impossible any termination of length. Without such a termination, there could be no use for it as a radius.

Turning to the second consideration, a straight line is not a curvature of zero. This zero cannot be equated with *not*. A straight line is a kind of line; it is not a kind of curve. The line is the genus, of which straight and curved are the most important species. In the next chapter, the reader will learn that a non-zero absolute can only be an indication on the same determinate system, as an angle of 0 is the beginning line from which an angle opens.

Let us close with another important aspect of the angle. The quantity of lines through a point cannot reach in every direction; this is because of the nature of an angle; a weather vane is not just incomplete; it is incomplete-able. In order to have more directions drawn, there must be more than one starting point. These additional beginnings are required in order to cover every possible direction. If a sphere were to exist with this starting point at the geometrical center, there could not be any more points than there are on the surface of that sphere. But, as was explained earlier, not all of these surface-points could simultaneously be intercepted by radii. If successive spheres were constructed around that, the same situation would exist in each case. Since lines around a point cannot be stacked, the radiating points would have to be staggered in such a way that the lines could cover every possible orientation. Nested spheres would not work for the same reason.

How could a system with lines pointing in every possible direction be attempted? Imagine the face of a cube. Along each of these innumerable points on each face, place neighboring parallel lines pointing perpendicularly outwards. The reason why the parallel lines must be next to each other is that no lines point in exactly the same direction; if they were some distance from each other, the spaces in between would have no direction indicators pointing at them.

But that would be not be enough. The edges would be left uncovered. There are twelve of them, each at ninety degrees from one another. To remedy this, along each edge a rectangular plate would need to be placed diagonally overhead touching the outmost parallel direction-lines stemming from each of the adjacent faces; on these diagonal plates, place rows of neighboring parallel direction lines facing outward. This would now require that the smaller twenty-four edges be covered. On each of these twenty-four edges, begin with the slanting plates fully equipped with direction lines; then on the forty-eight smaller plates, more direction lines, over and over again, exhausting the possibilities of man's power of enumeration. The calculations would be insuperable, the work impossible. Monkeys randomly typing letters might accidently come up with a Shakespeare play, but the greatest planners could not build this universal direction indicator.

Since we cannot know even so much as all the potential directions, space cannot be reduced to any number of man's observation points. It is not is just a projection of the human context. Space lies beyond all that man can do. A statement like the following from a finitist is wrong. "I want to start by stating unequivocally, there is no such thing such thing as 'space,' whether viewed as the infinite void of the Greek atomists, or the receptacle of Plato, or the absolute cosmic reference plane of Newton, or the acrobatic and curving frame of Einstein . . . . *There is no such entity.*"[133]

Of course, space is an entity. Neither material, nor mental, it is, nonetheless, an independent part of reality. It exists.

CHAPTER XI

ON THE DIVISION BY ZERO

Martin Heidegger famously asked, "Why is there something rather than nothing?" Whatever the final disposition of his question, the reader should understand that empty space is not absolutely nothing, that it possesses the attribute of homogeneity, yet its endless locations are, each of them, unique.

Let us now consider this question: When, if ever, is division by zero possible?

It is not possible in any case where a strict zero is meant. Simply put, there is nothing in zero with which to divide. As an excellent authority puts it: "division is a short method of making several subtractions of the same number."[134] Zero from any number is only that number; there is nothing in zero to subtract. Yet, people attempt to say $6/0 = \infty$. Take the alternative definition of division, that it is "the process of finding how many times one number is contained in another."[135] Quite simply, zero is not found in six. It is the absence of any quantity.

The famous theory of limits holds that since the quotient grows larger as the denominator gets smaller, it is permissible to think of ∞ as the limit of $6/x$ as "x" approaches zero. Yet, the fundamental fact remains that while a finite quantity can be divided into one part—that part then being equivalent to the whole—it cannot be divided into no parts; in other words, $6/1$ is possible, but not $6/0$. As was shown in Chapter VII, subtraction can use zero, but division cannot.

Under what conditions, if any, is anything resembling the fraction like $6/0 = \infty$ possible? The short answer is that it is possible only when zero is considered not as a symbol signifying an absolute nothing, but as a position on a scale or a number line.

In space, there is no zero absolute. Each position is a point, and this is independent of whether or not a human being has it located on some sort of a coordinate system. Since any such system, however fancy, must itself rest on a point, there is no getting around the fact of its dependency upon the prior existence of points. Each point in that infinity that is space is a position with neither length, width, nor depth.

The zero in a number line occupies a point, and that mark could have been placed in another position. The infinitesimal presently called "0" could have been marked "5". Yet, that position in space exists, whether it is marked "5," "0," or not marked at all.

Zero signifies both an absence of magnitude and an absence of direction. If the zero in a fraction like 6/0 is considered, not in terms of what it symbolizes— namely, as an absence of these two attributes—but as an infinitesimal occupying a position, then, since the infinitesimals in any finite interval are uncountable, the ratio between any finite quantity—6, 60, 600, etc.—and an infinitesimal is infinite in the sense of being immeasurable. The ratio between 6 and zero absolute is non-existent, but the ratio between six units on a number scale and the single infinitesimal in space which is designated as "0" is the lesser actual infinity, i.e., it is beyond any possible system of enumeration.

Strictly speaking, 6/$_\bullet$ is not a ratio, which is the measure of relation of one finite number to another. Since it involves a comparison of a finite number with an infinitesimal, it is a ratio only by courtesy. Such "ratios" are infinite, but not unlimited. They are immeasurable.

If one were to use it as a potential infinity, as is often the case, the symbol "∞" must not used as if it had an exact quantitative meaning. Otherwise, if 6/0 = ∞ and 5/0 = ∞, then 6 = 5—a plain impossibility. It is self-evident that the ratio between six units and an infinitesimal is greater than that of five units and an infinitesimal. (Neither should one forget the difference between potential infinity and actual infinity, as discussed in Chapter VIII.)

In the future, this type of *division* will be italicized in order to distinguish it from the ordinary kind. Division by zero is never possible; *division* by zero is possible, i.e, 6/0 cannot be, but 6/•, or under an alternative mode of expression in which the numerals are underlined, 6/0, can exist.

Considering just space, under what circumstances is this kind of operation possible?

The zero referred to in Cartesian representation is really a point. Although it signifies both an absence of quantity and an absence of direction, this is no different from any other point on a number line, such as +6, considered out of relation with anything else. It is important to understand that the location represented by this +6 is not the same as the interval, 0 to +6, which is finite. The point represented in Cartesian coordinates as +6, in contradistinction to the interval it terminates, is every bit as much an infinitesimal as that designated zero in that system. Although each point differs in space, they are all infinitesimals.

Put differently: In any drawing, that which is given the zero value is a representation of a point. It is not the point or infinitesimal itself, but it is the representation of it. The zero in any number line is also a point therein. To speak in terms of metaphysics, in order to have no being whatsoever, something would have to not exist; but even empty space exists. Therefore, since the zero in a geometrical figure represented in a drawing has at least location, it is an infinitesimal. Being an infinitesimal, the difference between this point marked "0" and a finite number must be beyond finite measure; for any finite interval, however small, contains immeasurable points or spacial infinitesimals. To return to the short definition of division, the process of subtracting an infinitesimal from a finite unit of length—such as six—is outside the realm of numbers.[136]

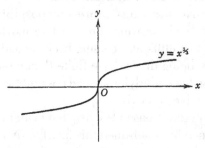

The figure on the left was copied from *Introduction To Calculus and Analysis* by Richard Courant and Fritz John. The figure is a Cartesian rendering of the function, y = f(x) = $x^{1/3}$, i.e, the inverse cube root of x. The curve is continuous for all values of x. But the derivative, y'= $1/3x^{-2/3}$, they say, would be infinite, and, therefore, cannot exist. At x = 0, it would involve both a division by zero and an infinite derivative.[137]

If the modern idea is correct, the notion of an infinite derivative would have to be rejected. A derivative like that would have to be classified as "undefined" at 0. Were this the case, then it would be a potential infinity.

Or is it to be treated as a *division* by zero, on the grounds that if the position occupied by zero exists as an infinitesimal, so must its derivative?

The resolution of this issue is as follows: in the case of a derivative which is based upon a certain refinement of the original equation, then an infinite derivative must be accepted. That certain refinement is a trigonometric tangent. And one of its angles is a right angle; at this angle, the ratio is a finite vertical divided by an infinitesimal horizontal. This tangent is like 6/•.

The tangent at 90° does not specify any length for the vertical, only that it is finite. The derivative at this point is off the chart, so to speak. The rate of change in this case is beyond numbers, past man's power of reckoning.

The term "undefined" is not exact. The derivative at x = 0, does exist, but in the realm of the infinite as immeasurable. It has a definition. In his standard work, *An Introduction To Logic*, H.W. B. Joseph lists six rules. "I. A *"definition must give the essence of that which is to be defined."*[138] In the course of this book, there is sufficient differentiation from that which it is not to avoid being confused with the (a) infinitesimal, (b)the finite, (c)potential infinity, and (d) that type of actual infinity that is endless. Furthermore, there are other kinds of immeasurable infinity besides a ninety degree tangent. This is enough to classify it as a derivative within the realm of that problem. "2. *A definition must be per genus et differentiam (sive differentias).*"[139] The genus is an actual infinity and the differentiam is that it be immeasurable, but not endless. "3. *A definition must be commensurate with that which is to be defined.*"[140] It is applicable to all that belongs to it. It is excluded from that which it is not, namely, finite and (where they exist) infinitesimal derivatives. "4. *A definition must not, directly or indirectly, define the subject by itself.*"[141] It does not. "5. *"A definition must not be in negative where it can be in positive terms.*"[142] It is not a negation; it is characterized as belonging to an actual infinity in the sense of being immeasurable. It is a positive. "6. *A definition should not be expressed in obscure or figurative language.*"[143] An infinite derivative in the infinitesimal calculus exists as an immeasurable infinity.

But, the reader should be cautious. It can hardly be overstated that this goes beyond mere drawings. Serious geometric drawings represent possible constructions in space, albeit imperfectly. Such drawings are not like movie cartoons of coyotes falling down thousand-foot cliffs and coming back seconds later with a few scratches. The lines really being drawn have finite thickness, but they enable the mind to fix attention upon the underlying spacial possibilities. Once that is granted, then an infinite derivative can exist.

It would be invalid if the point were material. Reason teaches that matter is finite. (For those who need it, modern science has confirmed this fact.)[144] In an actual physical machine, the zero mark could only be considered to refer to the infinitesimal in space occupied by the matter on which the mark was made, not the physical mark itself or anything else of finite thickness.

Consider a construction which begins as an asymptote, one with an equation like $y = k/x$. According to the traditional understanding, since line and curve do not meet in the finite, there can never be any value at $x = 0$. But we have seen that since the infinitesimals separating curve from line are not unlimited, even though the line may extend itself every time the curve drops closer to it, there must come a time when they must meet. The line might have to grow beyond $\mathfrak{I}_{\omega}$. But it would happen. However long that line might have to become, the tangent at 90° would exist. The definition is clear enough to be distinguished from anything else. The constructor of the curve may not expect to live to see it. But it is complete-able, there is a value which it would possess if it ever were completed.

Next take the exotic issue of 0/0. Were it is really nothing divided by nothing, the only possible answer would be nothing. Take, for example, the common $6 \cdot 0 = 0$; dividing both sides by zero, the result is $6 \bullet 0 \div 0 = 0/0$. If 0/0 equaled any number other than 0, the result would be absurd. If a person tried to say the two zeros cancelled each other, equaling one, then the result would be, $6 \bullet 0/0 = 0/0 = 1$ or $6 \bullet 1 = 1$, obviously a false answer.

Yet, $0/0 = 1$ in the form $\sin \theta/\theta = 1$ when $\theta = 0°$ is essential for obtaining the derivatives of trigonometric functions. In order to be valid, its parts could not be the same as in abstract arithmetic.

The result is inescapable. It is derived from a geometrical drawing which may be found in any calculus textbook. That drawing represents a possible construction of lines in space. From the drawing, the following inequality is obtained: $\sin\theta/\cos\theta > \theta > \sin\theta$. Dividing by $\sin\theta$ produces $1/\cos\theta > \theta/\sin\theta > 1$. When $\theta = 0°$, $\cos\theta = 1$ and $1/\cos\theta = 1$. Therefore, since $1 = 1$, $\theta/\sin\theta = 1$, or $0/0 = 1$.

Quite clearly, these are not ordinary numbers. If they were, that last division would be contradictory. What then are they? Consider first the numerator in the term, $\theta/\sin\theta$. Any θ, being an angle, is a relationship in space. These angles are usually measured in degrees. $0°$ is not an opening of the two sides of a vertex, but it is a terminal of that configuration. It is not the kind of zero in which 6 apples are subtracted from 6 apples. It registers the starting line of angle measure.

Second, let us consider the denominator, $\sin\theta$, which is equal to 0. This zero is the value of the sine of angle zero. It is not a complete nullity, one devoid of an necessary connection to function and angle. It is a position—one that is often calculated in such figures as sine waves. Contrast this with the following equation, $\sin 0° - \sin 0° = 0$; there, the subtraction is an emptying of being, the dis-establishment of any position whatsoever.

This 1 is not the 1 of unity with its rational subdivisions and its irrational parts. It is the *1* of identity. Hereafter, it will be distinguished from the usual form by being underlined. It is indivisible; it cannot be broken into fractions.

$0°$ divided by $\sin 0°$, like *division* by zero, is not an actual arithmetical operation. Unlike the latter, $0°/\sin 0°$ is rather a comparison of a peculiar angle with its sine, resulting in the *1* of identity. It is like •/• which, of course, must equal *1*.

Like 6/•, it applies not just to drawings about space, but to the three dimensions of actual space—to lines closed and enclosed and the points identified on those lines.

Aside from terms like 6/0 and 0/0, there are other forms, sometimes called "indeterminate." These are [A] $0 \bullet \infty = 0/(1/\infty) = \infty/(1/0)$; [B] $0^0 = e^{0\log 0}$; [C] $\infty - \infty = \log e^{\infty - \infty} = \log e^\infty/\log e^\infty$.[145]

[A], of course, is invalid when zero is used in an absolute way. In special cases, it is valid where 0 is in reality •. Where the finite alone is being considered, $0 \cdot \infty$ could be used to symbolize the infinitesimals within a line of increasing length, a potential infinity; the 0 referring to the absence of finite width in a position.

[B] is invalid when zero absolute is used as an exponent of 0; it would then mean that nothing is taken no times at all, not even once. Someone might attempt to argue that it is the one of identity. But that would be a mistake, since out of nothing, only nothing can come—a quite different case from the angle zero. The second term is also wrong in the same way, since zero absolute cannot have a logarithm.

If infinitesimals are involved, it would mean that (1) either a zero absolute has been taken an infinitesimal number of times or (2) that an infinitesimal is taken an infinitesimal number of times. If (1), the expression is without meaning, since *what is not*—zero absolute—is supposed to be multiplied by a position. If (2), it is impossible, since one cannot make the smallest any smaller. Moreover, a position times a position does not give a square position.

[C] is not generally valid. Its very first expression. $\infty - \infty$ is ambiguous. If it is used in the sense of endlessness, then it is a brain cracking contradiction; $\mathfrak{E} - \mathfrak{E} = 0$ would be the extinction of space, which is unthinkable. If it is not that, but refers to the difference between two potential infinities of the same kind, then it is a zero absolute. But the second and third terms to which it is to be equated are not possible, although "consistent" with what is usually taught. (This, the reader can verify at leisure.)

In summary, when zero is used to signify utter privation, division by zero is impossible. But when that zero is considered as only a point in space, it is possible to *divide* by • or *0*. The possibility exists because there is no such thing as a zero absolute in space. Every point is a position. It is not just that the infinitesimal occupies a point. It *is* that point.

What of the material object itself, as distinct from the space that it occupies and the time that lasts during any of these starts, stops, new starts, and finishes? To be sure, each part of it occupies the space of uncountable infinitesimals. Wouldn't the comparison of just one of those infinitesimals occupied by part of the object in comparison with the volume occupied by the whole object be infinite? It would indeed, but there might be apertures within the volume; it is possible that matter might have a granular structure; indeed, this is what the majority of opinion holds. The molecules of an object in a given volume of space are believed to be vibrating and, therefore, do not cover all the spaces at any given instant. We do not know whether the forces existing between the molecules are continuous or intermittent or, if the former, that they cover all the intervening space (which is improbable, anyway). Therefore, even if one began a measurement with matter at the border of an object, one would not know whether it was a continuous extent of matter from—say, 0 to +6 units. For that reason, one cannot speak of the continuity of matter with the assurance that one may speak of it for space and time.

But this does not prevent a division by *zero*. Even though we cannot speak of a continuous material extent of +6 units, we do know that matter must occupy a finite portion of that interval.

The question of division by zero with respect to time and motion will be discussed in the proper places later on in the exposition.

CHAPTER XII

INDUCTION

It is impossible to put a steel cube 2 inches on a side into a steel hole 1 inch on a side without breaking one or both. It is common to say that the hole is too small. It is equally valid to say that there is not enough space in the hole to put in the cube. A distinction must be made between that which is *spatial*, i.e, an object's volume, and that which is *spacial*, i.e., the space occupied by an object. Every physical object exists in space. It has to fit within it. Space has absolute reality.

The existence of the infinitesimal has been proved through reason. That having been accomplished, the fact that lines without finite breadth can exist also follows. Squares and circles can both exist, but a square circle cannot; it is impossible for this hybrid to exist in space.

Even if pure geometric figures can only be approximated by men, it does not make them phantasies. It is true that men cannot make them, but that is because the matter by which lines are drawn is by nature finite. Nonetheless, geometric lines remain possibilities capable of being realized in space, not mere products of consciousness. The fact that they can exist enables men to make approximations of them with lines that have thickness. The science of geometry is actual.

But they are more than possibilities. It is not just the science that is actual. Within the space taken up by the physical solid lines, there is room for innumerable lines without breadth. This fact allows us to use the thick lines as surrogates for the geometric ones. This is the case, not only with single lines, but in any well-constructed rendering of geometrical shape—whether triangles, rectangles, ovals, circles, etc.; this goes for solid geometry also.

Take the concrete triangle with the following sides: 3,4,5. If the representation is well made, then within its thick lines, there exist the actual lines which make the calculations sure.

Even crudely drawn lines can serve as a guide to help us visualize the drawing of these lines in space. The fact that we know that they can exist, even when not drawn correctly, is the basis for higher substitutions, such as those of algebra. Given the two sides, 3 and 4, we know that the remaining side must be 5. But we do not know this because of any arbitrary combination of letters in a formula. The triangle must be shown to have the property that the sides are commensurate to their squares.

Against this, it might be objected that if we look at a line under a microscope, we might see tiny places where the writing instrument did not make its mark.

To this, it is answered that in order for a straight line to get from the position at which it stopped to where its activity is resumed, the drawing device had to traverse the distance.

Against this, it might be asked, what if the writing instrument jumped up slightly? Then its actual line would not be the one intended. Does this not show that lines do not actually exist, that they are chimeras which men can never hope to attain, conceptions which can never exist in reality?

To this, the answer is that the jerks in the line might be in one place when applied and not there on a future effort. The intent was the same throughout. More important, what was intended was the traversal of a distance, and since this was accomplished, the possibility of a straight line through it remains. There is no need for it to be envisioned as a Platonic copy.

Against this, it might be objected that this defense cannot account for a curved line, which by definition does not follow the course of the distance between two points.

The answer is that since straight lines without breadth can exist in space, so can curved ones. If one unwraps a perimeter line to a circle, one has a straight line which is completely interchangeable with other straight lines; the same can be done in reverse. In solid geometry, a cylinder can be split and unrolled into a flat surface bounded by straight lines.

More generally stated: we do not have to go beyond perception in order to find straight and curved lines. They do indeed exist there. Even if there are breaks in the line, they are filled in. What we do perceive is that the line is unbroken *within our range*. Within that range, there are no places to be sutured tight. The brain is unable to discern the difference without assistance. This is the level at which geometric constructions are sensible. Whether the line is solid at the microscopic or sub-microscopic levels is not the issue. In either situation, what is perceived is a solid line. It does not require any breakthrough of the realm of the perceptual to form the abstraction of a straight line. The transition is straightforward. If the apertures are too small to sense, that is just the way they are.

Beyond that, human beings have an inborn capacity to perfect an attribute that they have apprehended. Being able to identify this lack for what it is—a deficiency—enables us to form a clear idea of that which does not have this imperfection. The mind proceeds from the realization that there is no reason why it could not exist in space to the understanding that it can. We have all used that power.

Suppose for the sake of the argument that the perfection men seek in lines could not possibly exist, even in space. Were that the case, then it would be the same as placing the steel cube two inches on a side into the one inch square steel hole. Geometry could not be a guide for science, for it would be in conflict with reality at a deep, deep level. Since neither engineering nor worthwhile physics could stand without such lines, geometry would be worthless as a guide for physical science. "Mickey Mouse" would do better; at least, students of the mind could glean something from the public interest in this fantasy.

In short, notions, such as Hume had, that mathematics concerns only relations between ideas and not between facts should be dismissed. Mathematical reasoning is based upon the facts of point and interval, and what is discussed therein can exist in space, even if men are unable to realize it because the lines they construct are only finite. What is decisive is the capacity of space itself to have the positions to receive them. The validity of perception requires the existence of the infinitesimal, which means that lines without finite breadth can exist.

The fact that these lines can exist in space is somewhat obscured when men talk about these figures existing on *planes*. But mathematical planes do not have an independent existence. They are abstracted from a practical condition of material existence, to wit: as a practical measure, to draw a figure, one must have a solid surface to support the thrust of the writing instrument. But this condition is not part of the figures depicted on the material surface as such; the properties of a circle or a triangle exist quite apart from that on which they are written. Their points are the same, regardless of what is used to support the drawing.

All the flat plane does is indicate that the points lie evenly upon it. It does not exist as part of the figure. It merely states a certain condition, as well as that the surface be hard enough to write upon.

Planes are like the construction lines in mechanical drawing; they merely assist without being part of what is being put down. With respect to the drawing itself, it does not matter whether the plane on which it is lying is an infinitesimal in thickness, or has the bulk of Mount Everest. The properties of the figure are independent of it. Although the figure's construction may require the assistance of the plane, once made it exists in space.

Since space is neither flat nor curved, all possible shapes and sizes can exist within it. A curved plane can be made. If it is convex, no sides of any triangle drawn upon it would be straight, and the angles would have to be greater than two right angles. It is not a film either; rather, it is the condition that the figures conform to a desired shape; that the points be comported a certain way.

The important fact about a plane in geometry is that it is an abstraction from material construction and is not conceptually a part of the figures themselves. A single layer of infinitesimals, all of them lying even, can exist in space; a curved surface can also exist. But the plane, as usually theorized, cannot, for it would take up no space. It is only an abstraction, summarizing an important condition.

The case of the plane shows once more the difference between figures which can exist in space and merely imaginary projections from practical experience. The figures have reality. They can be made from infinitesimals. The plane does not. It is merely the condition that the figures be presented without distortion on a surface hard enough to receive a material drawing.

Any type of energy, whether of matter or light, takes up space. Whether or not the thick lines drawn by men cover all the space within their boundaries, there is in them a similitude in all pertinent aspects between what can be and what is.

It is not necessary to interpret them as copies of Platonic pre-existent ideas. If the Platonic notion were true, it would actually be a kind of perception, a vision of the eternal idea of a straight line, suggested but not brought into existence by its rough material copy. This is not the case. *Science demonstrates the existence of the infinitesimal. This results, not from seeing another world, but from contemplating what is already here.*

As we shall see, the world of these geometrical figures and the world of material objects are the same. Both can exist in space. And this co-existence guarantees the validity of scientific induction.

For each of us, there was a first time that we observed a line that looks straight. A man-made object is not necessary. One could find them by looking at the sides of some far off-cliff—or even by creasing a leaf.

They are not just intentions going beyond a rough and uncertain basis. There is no contradiction in their existence, whether we are able to realize them or not. As such, they do not, as it were, conflict with physical objects. To the contrary, they complement them. Both are a part of reality.

Induction is reasoning from specific to general; deduction, reasoning from the general to the specific.

(This is not to be confused with what is termed "mathematical induction," which is, in general terms, a stair-step to the stars proof which will succeed with every number considered.)

Ayn Rand rightly discerned that the formation of a concept is very much a process of induction,[146] since it begins with a concrete and then makes an abstraction which would apply not only to itself, but to other things of the same kind. Once someone has abstracted the attribute of straightness, then he or she can refine it and deduce consequences from it. Connecting its definition with concepts at the same level or higher, as in the case of Proclus' improvement over Euclid, enriches it somewhat like a plexus of blood vessels, a bodily organ. By connecting the definition of a straight line not only with points, i.e., with infinitesimals, as did Euclid, but also with distance, Proclus had defined it with respect to two considerations more fundamental than a line.

But Rand was mistaken in supposing that in forming a concept, it is necessary that one must compare and contrast two or more objects, concentrating upon an underlying similarity amid their differences in order to form an abstraction. On the contrary, a single instance suffices.

As I wrote in an earlier work: "If two or more instances were required, a person could never arrive at a concept . . . , *unless they were presented simultaneously*. If one first perceived a single case of . . . phenomena, a shade of red, for instance, one could receive it in memory with sure accuracy. But how could one know enough to pull the earlier impression out of memory, unless one had already recognized it? The second shade is not required to enable us to see red as *red*. It can give us knowledge only of purity, intensity, and the like."[147]

Abstracting from a single instance and then retaining it in memory is an action of induction from a single instance. (Words assist, but they are not always necessary). Truths have been deduced from reflection upon a solitary attribute.

A universal is a fact of which there is at least one instance, but which is in principle repeatable. To illustrate: lines are universals, even the first line ever drawn or otherwise formed, by man or Anyone else. If there were never a second line drawn, it would still be a universal, since more than one is capable of being made. In a word, it would be repeatable.

Axioms or common facts are by definition universals which have been repeated often. Otherwise, they could not stand at the head of a process of deductive reasoning, which must presuppose them. Consider the axioms of Euclid. They were abstracted from everyday experience. Archimedes and thousands upon thousands have drawn from them. An example is Axiom 5, "the whole is greater than the part."[148] We have seen what happens when men try to deny it with Dedekind's definition of infinity as that in which the whole can be put in one-to-one correspondence with a part of it.

There is a third process which is neither induction nor deduction, but intimately related to both. It is a new finding of reason. An example is the discovery of the irrational numbers. Deductively, it was found that the third side of an isosceles right triangle could not fit into whole numbers and fractions. Yet, the geometer knew it had to exist. Inductively, this single instance was enough to raise to an abstraction and define as a type of number, the magnitude of which could not fully be comprehended by the nature of a unit. Perhaps, the original discoverer of this basic fact of nature had to check his demonstration over to make sure that he was not dreaming, but this peculiarity of that triangle was enough to convince the mind.

But before that abstraction could be made and raised to the level of an induction, something had to be recognized. This was that another kind of magnitude besides whole numbers and fractions must exist. To deny its existence would be to deny both past perceptions and reasoning on perception, i.e., the isosceles right triangle and the nature of a unit. This realization was neither reasoning from general to specific nor from specific to general. This third element is discovery by reason alone.

This discovery was a breakthrough. Yet, it was not perception that had been broken through. What was denied was the conceit of the Pythagoreans that numbers are everything, and that when men place their marks on the two endpoints of a magnitude, they can precisely calculate every part of it through the selection of a unit.

The discovery of this fact was humiliating to human pride. It is reported that the Pythagorean brotherhood put the original discoverer to death. Something like that brotherhood exists today. We find that in the modern attempt to define the incommensurable by means of the commensurable, discussed earlier.

Another example is the discovery of the infinitesimal. To disclaim it is to repudiate sense perception. This is also the case with the discovery of the veritable number, a subject to be discussed in the next chapter.

The transition from the recognition of room to the identification of space is not such a breakthrough. There is no resolution of a conflict between reason and expectation, as with $\sqrt{2}$, nor is there a requirement that the inherently imperceptible be admitted as a logical requirement of the world disclosed by perception. Rather, it is a natural and inevitable result of further abstraction from a fact of reality understood by men and by such animals as cats and dogs.

That both the so-called empirical sciences and geometry and, therefore, arithmetic, refer to the same reality is sometimes obscured by the greater certainty of the latter two. The geometer is quite easily able to determine that plane triangles are either equilateral, isosceles, or scalene. But the empirical scientist has more trouble. Suppose that having found that after having tested several types of A that A is B in every case: it would only be reasonable for the scientist to conjecture that all A is B. But he still cannot universalize his discoveries: it may be that there is a type of A which is not a B. He knows that he has not identified all the kinds of A that there are. He may have a belief that when he has explored all sides of A he will apprehend the proposition, All A is B, to be true. But he still cannot be sure. This has led to the wide-spread belief that mathematical truths are very different from those of empirical science. If they are really the same, why is it, as John Stuart Mill once remarked, that in some cases a single instance will prove a something to be a fact, and in other cases, hundreds of observations are not enough?

This problem was basically solved by John Cook Wilson. Whether others preceded him I do not know. His words on that subject are scattered throughout his *Statement And Inference*. The solution was brought out in Chapter VI, but because of its crucial importance for the unity of the sciences, it needs to be stated again in a different way. The following is my own adaptation of his idea, which appeared in *The Stance Of Atlas*:

"With simple things like the lines and curves constructed in plane geometry and elementary arithmetic, algebra, and analytical geometry, the necessities are apprehended easily. In plane and solid geometry, for instance, the properties of points and lines are easily grasped. The attributes are so simple that quite complicated demonstrations involving them can be followed with ease. The truths are self-evident. In slightly more complicated areas like calculus, the answers obtained are obviously true, although there is some debate among mathematicians over the foundations; this last is due to the intellectual disorder in this century.

"Much, much greater difficulties, of course, come in ordinary life. There, so many sides exist that many spend most of their time trying to account for it. This is one of the chief reasons for the seemingly endless number of stories about love between the two sexes, each author taking some tangent and exploring the consequences.

"To a marked extent, this is also the case with the natural sciences (although they do not have free will to contend with). In plane geometry and solid geometry, a single demonstration is sufficient. Once we have proved the Pythagorean theorem, we need not do so again with a different triangle. It is not necessary to remember the proof; only that we have proved it once. But when one is studying the relationship between the green color of a plant and the role that light plays in vegetable life, many observations from many samples are necessary to convince oneself that it is not just an accidental association. Many more are required to find exactly what the connections are. When one is dealing with matters which are too small or too large to be seen with the naked eye or which cannot adequately be explored through touch, the possibilities of making a false observation are immense. Then there is the problem of determining whether all the sides of the phenomena under consideration have been found. In some situations, there may never be a complete apprehension. If that be true, then probability is all there can be in these cases."[149]

In a nutshell, this is the answer to the problem of induction. When one studies a geometrical figure, every part of it is present to us. In the physical sciences, frequently, we do not know in advance whether this is or is not the case. We cannot pre-determine how many sides or aspects to a problem there are. We have found that A and B cause C with such types as A_1 and B_1 and also A_1 and B_2, ditto A_2 and B_2, the same for A_3 and B_1, etc. But whether all the types of A and B are basically the same and will result in the production of some type of C we cannot know until we have gone through all the types. And this determination cannot be realized until and unless we have exhausted the phenomenon.

It is not a question of the number of instances, *per se*. Were the individual instances all of the same quality, it would not matter whether one or a million repetitions of the same experiment were made. If the experiment were reported correctly, any additional trials would only repeat the same results. The re-performed experiment would merely prove to the second scientist that the first one had done his job. Third parties would also be given more assurance, since it would then become less likely that the first scientist had reported some falsely. But the situation is not so different with geometry and arithmetic in that respect. There, too, the person who reads the account of a proof may need to check it. And the circumstances of the third party who takes what the other two had claimed for granted is similar to that in the physical sciences.

The inductive establishment of a basic abstraction in good mathematics is easier, because its validity can be checked by anyone with a sound mind and sufficient learning. Underlying both types of investigation, however, is the sameness of reason. An ornithologist of a couple of centuries ago who was studying swans would not characterize their feathers as being of "n" colors, if his only experience was that of white. After the black swan was discovered in Australia, he then could speak of its being either one or the other. Only if the scientist had specific knowledge of the appropriate part of the bird's DNA and related inner structures could he determine whether or not other colors were possible in nature. The same is the case with a mathematician who speaks of space having "n" dimensions, meaning some number greater than three when he cannot show through induction or prove through demonstration the existence of that plenitude. Merely to suppose something to exist because it is consistent with some fashionable formula or turn of thought is bad evidence. Doing that is not opening oneself up to new horizons, but obscuring what one does have.

Everything is in space, and no material object of any kind can go out of space. The science of pure space is geometry. Its conclusions are certain, and material bodies must be coherent with that. From spacial considerations come lines; and from that, measurement. Even the similar objects supposedly used by early men when they were acquiring knowledge of counting were distinguished from one another by their positions in space. A science of numbers that tries to ignore this connection is not more abstract, merely disconnected. (The questions of time and motion will be brought up later, but, as we shall see, they are not in conflict with what is being said here.)

When mathematics is structurally geometrical, it not only touches base with the materialistic experimental sciences, but is, so to speak, in contact with them everywhere. Both bespeak the same reality and can rely on each other. Recognizing that both geometry and physical facts apply to the same reality provides a solid basis for intellectual advancement. Sometimes one is asserted to be above the other. In the 17th century, Newton was able to show through geometry that the area of a closed curve, such as a circle or an ellipse, could not be expressed by any algebraic equation; many, including Leibniz, disputed this, but the doubting analysts finally admitted defeat a couple of centuries later.[150] In the century just past, the tendency has been to affirm the superiority of analysis over both algebra and geometry.

An example was Felix Klein who held that the idea of symmetry in group theory was higher than these.[151] This, however, is not true. Consider an empty plane: to try to express it in terms of some rotation or substitution of a group would involve great perplexity. Look at the equation, y = x: in geometry, it is two lines of equal length; in coordinate geometry, is a 45° line. How could a sophisticated group theory express this more "fundamentally"? Look at y = x +2! Lines and planes are more basic than groups.

Typically, "groups" involve cycles and replacements. Both presuppose space and motion. To attempt to substitute space and motion with mere symbols is nothing less than an attempt to evict reality. The very intent to get away with it cannot be expressed without reference to spacial or motional terms. What do "get away" and "with" mean anyway? The idea behind throwing away the ladder after one has climbed up to a certain height can only be to prevent the Wittgensteinians from knowing from whence they came and, therefore, from realizing where they are. Cortez burned his ships so that he and his men would either fight or die. These moderns are trying something less human.

A symbol is the use of one existent to stand for one or more other existents.

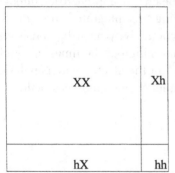

A geometrical drawing is an example: lines with finite thickness are used to represent those without such thickness. In full symbolism, where the recourse to the senses is less, the intellectual should occasionally refer back to the basis from which his thought originated. He must be sure that the various extensions he is making are valid. Mere analogy is not enough. For instance, conflating potential and actual infinity will lead to the idea that dreaming is the same as actual construction.

Both geometry and more abstract symbolism can be valid, but reality surely comprehends both. In a tiny respect, Hilbert was slightly right in suggesting that some axioms can apply both to beer mugs and the figures of geometry; both take up space.

Consider the familiar expansion of $(X + h)^2 = XX + Xh + hX + hh$. This binomial has a concrete correlate, as all will recall. X and h are both lengths. Each term is a rectangular product of two of these lengths. It is clear from this that hh is as much a part of the expansion as are the other terms—that it is as necessary as the other parts; if it were to disappear, so would Xh and hX, leaving only XX. This follows from the nature of the squaring of a binomial.

What if some physical phenomenon expressible by the expansion of $(X + h)^2$ were represented without the term hh? The reasonable person would think that either there was something else which concealed its existence, or that a mistake was made. This topic is the theme of Chapter XVIII.

Henri Lebesgue once wrote that "by the use of coordinates [Descartes] reduced all geometries to that of the straight line, and that the straight line, in giving us the notions of continuity and irrational number, has permitted algebra to attain its present scope."[152] It would be more accurate to say that Descartes gave us two mutually perpendicular straight lines as the determinants in terms of which curves capable of existing in the area between those two lines could be recognized. Without that area, his coordinates and their equations would mean very little. (Descartes' geometry does even not exhaust all the geometry that was known by the ancients.)

In summary, not only physical objects and light and radiation, but also geometric lines can exist in space. It is the fact that both the subject matter of physical science and geometry exist in space that makes the latter such a wonderful guide to the understanding of the foregoing. Geometrical ideas are induced from the attributes of experience and conceptualized into definitions, axioms, and premises for subsequence reasoning. The major difference between the realization of inductive truths in geometry and in empirical science is that the ascertainment of truth in the latter is much more complicated and often concerns things which are either above or below the level of apprehension through direct experience. Since counting and measuring are integrated through the number line, the results of algebraic and analytic activity must be consistent with its properties. On account of the intellectual milieu of the times in which men live, either the geometrical, on the one hand, or the algebraic and analytic, on the other, is usually given preference. The number line embraces both.

CHAPTER XIII

THE VERITABLE NUMBER SYSTEM—

THE SOLUTION TO THE MYSTERY OF √-1

√-1 is a great paradox. It is a logical impossibility, if derived from what is called the "real number system." In that system, the product of two negative numbers is always positive; -1 x -1 = +1. A negative product can only be the result of a negative number times a positive number; -1 = +1 x -1. Such a product cannot be a square root, since that is by definition a number multiplied by itself.

The square root of negative one has been dubbed an "imaginary number." Despite the scandal of being contradictory to the number system out of which it has supposedly been generated, it has been used with increasing assurance since the 17th century. In 1777, the Swiss mathematician, Leonard Euler, invented the symbol "i" for it.[153] Electrical engineering would be something very different without it. At least one famed historian has accounted it to be a hallmark of Western culture.[154]

Einstein made use of it in both his special and general theories of relativity.[155] Crucial to his notion that time and space are not independent, but conjoined as "space-time," is the stipulation that the time expression involves a √-1. This imaginary quantity is supposed to stand for an ineffable "fourth dimension" we humans cannot visualize.

The 19th century founder of symbolic logic, the Englishman, George Boole, characterized this number as an "uninterpretable symbol."[156] Yet, a practical interpretation existed before his time. It is said to have been first published by the Norwegian engineer, Caspar Wessell, in 1799. He showed that the imaginary number axis may be found at right angles counterclockwise with respect to the positive pole of the real axis.[157] (See figure i below).

One can only marvel at the intuition of those great mathematicians who did battle on its behalf, despite the fact that it was in plain contradiction to the kind of numbers which they understood. They knew that it exists without knowing what it is. Mathematics, to them, was not a matter of deductions from arbitrary stipulations, but an expression of fundamental truths. What they saw by intuition will shortly be made clear.

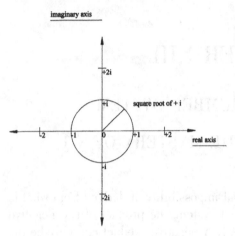

On the left is the famous real-imaginary axis. The real axis is horizontal with its positive portion right of zero. The imaginary axis is set vertically at right angles in a counter-clockwise direction from the real axis. The arrow indicates the positive portion of its axis. (The significance of the circle will be explained later.)

The positive part of the imaginary axis was erected through a counterclockwise direction from the positive real axis. It stands at right angles to the real axis. The negative i's indicate a clockwise erection at right angles from the positive real axis.

The unit circle clearly shows that real numbers +1 and -1 have the same absolute value. What is surprising is that +i also has the same absolute value as the other two, despite the fact that it clearly is not a real number. (Why this is true will be shown a little later.)

What also needs to be mentioned is the important fact that i does not equal $\sqrt{-1}$, but $\pm\sqrt{-1}$.

This brings up an interesting question. Since this particular square root is already negative, how can it be ever be positive? (The answer to this question will be apparent later.)

Everyone is familiar with the common form of multiplication, in which a length and a width are multiplied to produce an area, or in a table with rows and columns. This type is to be called "rectangular multiplication." In the calculation of imaginary values, a different type of multiplication, dubbed here, "rotational multiplication," is used. (In the next chapter, a third type will be introduced.)

Let us now consider rotational multiplication. Crucial to it is the unit circle. This can be seen through the famous De Moivre formula: $(\cos\theta + i\sin\theta) = x$. When the angle $\theta = 0°$, the real value is +1 and the value of the imaginary is 0. From there begins the rotation. At the vertical, $\theta = \pi/2$. There, the imaginary value is +i and the real value is 0. Ninety degrees further in the counterclockwise rotation, $\theta = \pi$ and the real value is -1. In the intervals between 0 - $\pi/2$ and $\pi/2$ - π, the values are part real and part imaginary; for instance, in the first quadrant, the real part progressively goes to zero while the imaginary part starts at zero and progressively goes to $+i$. Yet, throughout the entire range from $+1$ to $+i$ and from $+i$ through -1, the absolute value of the radius vector is the same. A similar result occurs with the Eulerian exponential, $e^{i\pi} = -1$.

The absolute value of the radius remains the same, irrespective of its position. Look at the line indicating the square root of $+\sqrt{i}$. This is shown to be the complex number of the form $\mathbf{a} + \mathbf{ib}$ where $\mathbf{a}$, the real part equals $(+\sqrt{2}/2)$ and the imaginary part, $\mathbf{ib}$, equals $(+\sqrt{2}/2)i$. The radius vector representing this number exists at 45° counterclockwise from the positive real axis. It can be found through the Pythagorean formula of the square root of the sum of the squares of the absolute values of the coefficients of its two parts, the real and the imaginary. This value is, of course, $|1|$—significantly, the same absolute value as its square, $+i$.

Now, let us examine $(+\sqrt{i})^2$, this being an uncommon way of stating the prototypical imaginary number itself, i.e., $(+\sqrt{i} \times \sqrt{i} = +i.)$ Wessell discovered that in this kind of multiplication, the two rotary angles of two consecutive vectors are added and their magnitudes are multiplied. A similar analysis was performed by the Frenchman, Jean-Robert Argand.

By this logic, an additional rotation of ninety degrees counterclockwise from $+i$ is the equivalent of multiplying $+i$ times itself, i.e., $(+i) \cdot (+i) = +i^2 = -1$. In the plane formed by these two number axes, doubling the angle is the equivalent of squaring a value. A similar operation takes place clockwise from $-i$ to -1.

Furthermore, these crosses with the imaginary axis can generate $+1$ as well as -1. This can be shown through a modification of the Eulerian formula itself. If $e^{i\pi} = -1$, using an exponent with 2π in $e^{i2\pi}$ produces $+1$; this should not be too surprising, inasmuch as this doubling of the exponent completes the full circle. It will be recalled that since the unit circle began with positive one, it could not do otherwise than end with it. Thus the unit circle can make a full revolution like a wheel.

Understanding rotational multiplication is crucial for handling imaginary numbers. It is reported that even the great Euler made a mistake here. He argued that $i\sqrt{-4} = 2$, his grounds being that $\sqrt{[(-1)(-4)]} = \sqrt{4}$.[158] That would be correct in rectangular multiplication, but not there. The correct answer is (-2).

It is important to note that the imaginary system is defined in contradistinction to the real system through the number line. It is by means of its being represented geometrically at right angles to the real number line that the early investigators were able to make sense out of it. It was not defined by free constructs, divested of all geometrical significance. Once more, the number line is proven to be indispensable. Without the right angle between those two number lines, the connection between the two would be unintelligible. The idea expressed that modern mathematicians have emancipated numbers from that line[159] is refuted even on the basis of their own practice.

This is a brief statement of how the square of imaginary one has, historically, been shown to be a real negative one.[160] As Boole pointed out, we proceed from interpretable expressions through uninterpretable expressions and then, finally into interpretable expressions.[161]

But there are still some unanswered questions, the most important of which are these: Whence came this expression that is un-interpretable within the real number system? Is the -1 inside the radical sign just the real minus one found on the horizontal axis in the second quadrant, or does it symbolize something else?

In the 20th century, it was standard practice to regard such questions as a little silly. To professional mathematicians like W.W. Sawyer at Wesleyan University, the square root of a negative number was actually nothing more than an instruction to perform operations "which *look* like equations for numbers, and might easily be mistaken for numbers," but are not.[162]

But this supposes that the "-1" inside the square root sign must be a real number, the inclusion of which cannot, therefore, be taken literally. And this, as we shall see, is false.

Let us begin by inducing the rudiments of the number system in which the square roots of negative numbers do make sense outside of rotational multiplication. In this system, -2 times -2 equals -4 and, of course, the square root of negative one times itself always equals negative one. Let us set the two systems side by side:

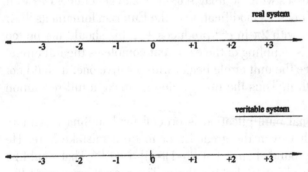

The name "veritable" for the new system was chosen after Gauss, who said that a number system behind the imaginary would be a "veritable shade of a shade."[163] But, as the reader will see, it is capable of sheltering *i* from the heat of contradiction and a lot more.

Let us note three essential similarities between the two systems and one crucial difference:

(1) Both employ the number line to its fullest. Both, therefore, are continuous. There are no extra points possible for the square roots of negative numbers not found in the real system. In the veritable system, $\sqrt{-1} = -1$. What of the other negative square roots? Consider $\sqrt{-5}$. This number corresponds to a negative number in the real system

which is the negative image of the corresponding irrational number on the positive real axis. For example, an approximation of the real $\sqrt{5}$ is +2.3460679775. Corresponding to it on the real negative axis is -2.346069775. The square of this irrational real number, however, cannot be a negative real number; it can only be a real positive five. But on the veritable axis, it is -5. Since a place exists for +5 and -5 on both axes, no new number is required in either place.

(2) Absolute values, or numbers without sign, exist in both systems. In each system, for instance, the absolute value of i is $|1|$. Similarly, the absolute value of the complex number $\sqrt{i}$ is also $|1|$. Once the magnitude of the unit has been established, it is the same for both systems.

(3) Zero is a legitimate value in both systems. Under either system, if -4 is subtracted from -4, the answer must be 0; also, on the positive side of an axis, +4 minus +4 is 0.

But there is an essential difference between the two: in the veritable system, positive numbers cannot be used together with negative numbers in the operations of addition, subtraction, multiplication, division, and root extraction, and the setting up of ratios. That is why that divider in Figure ii is shown at zero. Not every function which is possible in the real number system is possible in the veritable system, and *vice versa*. To understand what this means, recall that in the real system, one can multiply +2 and -3 and get -6. This property of numbers doubling back or reversing the direction of their sign, is to be designated as "reflectivity". Because of this property we can perform the following real operation: +1 x -1 = -1. For the same reason, the quotient of two negative numbers can only be a positive number, in this instance: -1/-1 = +1. But in the veritable system, reflectivity is absent; positive numbers cannot be multiplied and divided with the negative numbers: there, -1 · -1 = -1, which necessitates the corresponding division: -1/-1 = -1. It follows that -1 + 1 must be impossible for the same reason that -1 · -1 = -1. *Otherwise the effect of the second -1 would be the same as +1; then, neither symbol would represent direction, making both the same as absolute numbers.*

Since multiplication and division are repeated acts of addition and subtraction, there can be no reflectivity in those operations, either. In the real system, (+1) - (-1) = (+2), and the act of subtracting a negative number from a positive number of the same magnitude reverses direction. Because of this, the following addition is necessitated: (+2) + (-1) = (+1). Euler once justified the equilibration of subtracting a negative number to adding a positive number on the grounds that "to cancel a debt signifies the same as giving a gift."[164] In the veritable system, by contrast, neither of these operations is possible: there can be no reflectivity. All operations must take place completely on one side or the other of "0".

This idea may be thought of as somewhat strange. But consider: At one time, even the use of negative numbers themselves was disallowed. Thoughtful mathematicians asked: how could one subtract 5 from 4 and end up with less than nothing? And of course, they were right as far as they went; nothing can be taken away from nothing. Yet, it came to be seen that numbers can have a directional aspect. We take for granted the notion that right of zero all the numbers are positive and left of zero they are all negative. Veritable numbers resemble the original system in that no addition, subtraction, multiplication, division, or extraction of roots can cross over zero; but, like the real system, they are directional.

It is important that the nature of signs be considered a little more deeply, since it is only with respect to this issue that the real and veritable systems differ. In the real system, '+' and '-' signify both the operations of subtraction and addition, on the one hand, and direction on the other. To prevent confusion, the character "~" will be used to signify subtraction in the veritable system; this symbol was originally proposed in the 16th century. To signify addition in the veritable system, the symbol "&" will be used. Both can be found on the standard computer keyboard. The usual symbols for multiplication and ratio will be observed. "√" will stand for root.

Compare the two number systems: As long as the signs are the same and the subtrahend is smaller than, or at least equal to the minuend, addition and subtraction are the same in the two systems. Here are the results in the real number system. **A.** (-11) + (-7) = -18; **B.** (-11) - (-7) = -4; **C.** (-11) - (-4) = -7. Here are parallel results in the veritable system: **A'** (-11) & (-7) = -18; **B'** (-11) ~ (-7) = -4; **C'** (-11) ~ (-4) = -7.

A remarkable contrast is how they each handle absolute numbers. Today, the original system with no negative numbers is preserved in the form of absolute numbers, indicated by double vertical bars, one placed on each side of the requisite numbers, for instance, |3| or |.3|. In this adjunct to the real system, these numbers have the same significance as positive numbers. The reason why they do not have a like significance for negative numbers is that the latter are considered to represent smaller quantities as well as opposite directions; in fact, the greater the negative number, the smaller it actually is in that system—this despite the fact that the intervals between corresponding numbers on the negative and positive sides of zero are the same, i.e., +5 is greater than -7, even though the magnitude of the latter is greater. Negative numbers are considered to be inferior; this can be seen from the fact that in *order to move from the negative to the positive one must add; conversely, to move from the positive to the negative side, one must subtract, i.e., the quantity must be reduced.*

In the veritable number system, by contrast, negative and positive signify direction only, never size. In that system, -a > 0. Any number added to negative one in the veritable system can only be one of greater negativity. The total veritable operation is like the central axis of a double cone extending outward; nothing from one side can interact with the other. If two horizontal branches of the hyperbola face each other across a focus that is 0, can the right side be any greater than the other?

The symbolism for absolute numbers is identical in the two number systems; but, since absolute numbers have no signature, they cannot be used in the veritable system the same way as with the real. There being no predisposition in favor of the positive sign, an absolute number used in the same operation with a negative number must be treated as another negative, and an absolute number used with a positive number would be positive. $|7| \sim (-11)$ has no answer in the veritable system, because the subtrahend is too large, but $|7| \& (-11) = -18$.

To repeat: reflective operations cannot be done with veritable numbers. 3 cannot be subtracted from 2 to get -1. However, although you cannot go across zero, you can go through it. You can make the following argument: $(+2) \sim (+2)\& (-1) = -1$. This is because $(+2) \sim (+2) = 0$ and $0 \& (-1) = -1$, zero being common to both the positive and negative sides of the veritable number line. The answer is the same in both magnitude and direction as that given in real numbers, but the procedure has to be different because the underlying identities are different. The two must never be confused. In this regard, note that $0 \sim (-1)$ is impossible in veritable numbers but $0 - (-1) = +1$ is possible in real numbers. A negative veritable number is always greater than 0.

Yet, although it is impossible to combine negative and positive veritable numbers with each other through the five fundamental arithmetical operations, imaginary numbers can be multiplied and divided with real numbers. This can be seen in the equation, $e^{i\pi} = -1$. Imaginary numbers have characteristics of both the real and the veritable system. They are a hybrid and contain the square root of negative one and yet they are reflective. How is this possible?

To understand how this can be, let us first note a peculiarity of imaginary numbers. It will be recalled that i is to be defined as plus or minus the square root of the real negative one, not just as $\sqrt{-1}$. This reflects its dual nature. Neither is this the equivalent of +1. $\sqrt{-1}$ means that this veritable number has a real coefficient of positive one. $+i$ stands at right angles counterclockwise with respect to +1. It cannot embody +1 as part of its definition; the only use such a number can have is in the reiterative $+1i = i$. The De Moivre formula clearly shows that there cannot be any trace of a real magnitude inside positive or negative i. On the route from +1 to $+i$, for instance, the real component is gradually squeezed out.

But there are the signs. "+" and "-" are the part of the real that is always present in whatever quadrant the real number may be found.

The base comes from the veritable system and the coefficient comes from the real system. The composite character of imaginary numbers clearly shows in the case of -2i; if the whole expression were all in the real number system,

$-2\sqrt{-1}$ would clearly be a positive number; in that system, the product of two negatives has to be a positive.

Note that along the horizontal axis, it is possible to express the square of any real number, such as 2×2 or $2^2 = 4$. In the next chapter, it will be shown that it is impossible to express i^2 on the imaginary axis. This is because of the peculiar nature of the veritable component of imaginary numbers.

How can there be a real sign without any real magnitude? The answer lies in the practical interpretation of Wessell and Argand. It is defined as existing at right angles counterclockwise from the positive portion of the real axis. This counterclockwise direction is defined as positive. Correspondingly, $-i$ is defined as existing at right angles from the positive real axis in a clockwise direction. Once the choice is made as to which direction is which, no alternative exists. At zero degrees, the imaginary is 0 and the real is $+1$. At $\pi/2$, the imaginary is at its full value of $+i = +\sqrt{-1}$ and the real is 0.

Additions, subtractions, multiplications, and divisions between and among imaginary numbers of both signs are possible. The reason is that each imaginary unit has the same veritable content, the square root of negative one; there is no difference between a particular imaginary number and another imaginary number with respect to having the same veritable content of $\sqrt{-1}$. The difference is only in the real coefficients. (This result is fully discussed in Chapter XIV.)

Each quadrant of the unit circle of the real-imaginary plane determined by the two axes shown in the first figure constitutes a field. Every integral, fractional, or irrational value between $0 - \pm1$, between $0 - \pm i$, and every complex number formed from them has the potential of being formed in that field.

The plane consisting of these four quadrants and their respective fields extends beyond the unit circle. By virtue of the same ordering that establishes $+i$ and $-i$, the serial replication of the positive and negative i's is established along both parts of the imaginary axis. The next integral levels above the unit circle are $+2i$ and $-2i$.

In the days before negative numbers were accepted by mathematicians, absolute numbers were all that most experts would accept. With the introduction of negative numbers, the two pure systems became possible; first the real system was actualized, and now the veritable number system is in its inchoate form. It must be emphasized that neither the real nor the veritable numbers were derived from the other. Each results from a fundamental possibility inherent in the nature of negative numbers, namely the presence or absence of the property of reflectivity.

Knowledge of the existence of the veritable number was not deduced from anything. Neither was it arbitrarily assumed to exist. Nor was it established by a comparison and contrast between it and some other negative square root, like $\sqrt{-3}$. Rather, it was induced from a single case. What was seen in a flash was the rule in terms of which the existence of such a number was explicable.

And now, we are ready to consider the mystery of the -1 within the radical sign. Is it a real number, or is it a veritable number? Stated more emphatically, is it only the square root of minus one on the negative portion of a real axis, or does it indicate the veritable portion of an imaginary number? This is a question of some importance, since negative square roots presented themselves as solutions to algebraic equations centuries before the geometrical solution was found.

The answer is that it is veritable. If it were real, then, as was stated before, a term like $-4\sqrt{-1}$ would be $+4$, the reason being that the product of the two negative signs would be a positive. The difference does not lie in the magnitude of the numbers; in the representation of the complex plane, the units are the same size. The difference is in the way that two different kinds of numbers use their signs.

By having the same magnitude as the real number, the imaginary-Y axis is just as significant as the real-X axis. If they were of different sizes, instead of a unit circle, there would be a unit ellipse. Euler's formula would not hold.

The veritable number system was discovered on account of its ability to produce square roots of negative numbers. It is the necessary alternative to the real number system.

That is why there was still an element of mystery, even though the essence of rotational multiplication has been understood for a couple of centuries. Having overlooked the fact that $\sqrt{-1}$ is a length, they leaped to the conclusion that it was the square root of a real -1. They missed the fact that there was a rotation between one kind of line to another. Twentieth century theorists supposed that it was just an impossible real number symbolizing the operation of reversing direction; having made this assumption, they then performed calculations which always worked anyway.

But it is not what W.W. Sawyer called an "operation." The Y axis, although "imaginary," has an identity of its own. Its numbers not only look like numbers; they *are* numbers, but of a special kind. The actuality of the length of the units on the Y-axis cannot be denied. If they were of a length different from the units in the X axis, the unit circle could not exit. There is no magnitude in the mark of a sign alone; standing by itself, it only indicates on which side of "0" it stands.

But neither does the Y coordinate bring about rotation all by itself. As the radius vector alters its position in the unit circle, the values of its real and veritable components change. There, the real X component is as much an "operator" as $\sqrt{-1}$. Yet, no one would deny that X units are just plain numbers and not instructions which merely resemble numbers.

The real numbers of the X coordinate and the veritable numbers of the Y coordinate vary across the coordinates. At +1, the value is all real, none imaginary; at +i, the reverse is the case; and at -1, the veritable element has been squeezed out, all of it being replaced by the real. This alternation could not take place if the two kinds of numbers were not distinctly different in nature; rotational multiplication, for instance, would be impossible with crossed positive and negative real axes. The difference between the real and the imaginary axis is not in the signs, which are real in both cases. What distinguishes them is that the numbers in the X axis are all real while the reiterative core of the imaginary Y axis is veritable. The rotation in rotational multiplication can take place because of this alteration in which the veritable and real variables act like innumerable cogs, serving as ever changing pairs in an exotic clock.

To recapitulate, at the level of the infinitesimal, direction cannot be discerned. But once the level of the finite is reached, straight lines exhibit singular directions. Units of the same length differ both because of orientation and because of the way that orientation is handled. The absolute line is not lacking in that attribute; it can point any way. But the line divided by zero has two opposing directions, and the way they are considered makes all the difference between the veritable and the real.

With the discovery of the veritable, a new number line has been added to the absolute, the real, and the imaginary. Moreover, it is more complete than the real, since positions can be named on it which cannot be designated in a straightforward manner on the real. Because of this, it is probably a better model for the number line than the real; the cautionary word, "probably," is used because the veritable system is a recent discovery which still requires exploration.

We can only marvel at the men who knew there was a fundamental connection between this strange square root and the real number system, despite its seeming contradiction. Euler, for instance, understood that the logarithm of a negative number would have to be imaginary[165]—which means that the real number system is not complete in and of itself, but requires supplementation.

This brings up another question: How is it possible for the real and the veritable systems to combine into the imaginary system, even though the two are incompatible? This is the subject of the next chapter.

CHAPTER XIV

FROM THE VERITABLE TO THE IMAGINARY

With the discovery of veritable numbers, the mystery of $\sqrt{-1}$ was solved. Two questions will be considered in this chapter. The first is the compatibility of the veritable line with the real line. The second is the more complicated question about how imaginary lines are possible.

Consider whether a real line and a veritable line can be employed mutually as perpendicular Cartesian axes. It might be argued that since the real line would only produce horizontal values and the veritable line, vertical values, that there would be no conflict, since the coordinates of any given point (r,v) would not by themselves use any of the four forbidden operations.

That much is true. There is no contradiction in a point being coordinated by mutually perpendicular axes. Now, consider what would happen with two neighboring points. If they were just counted as point (r_m, v_n) and point (r_{m+1}, v_{n+1}) or just in absolute numbers as 2 points, there is no difficulty. But if a combinatory operation of any other kind were attempted, there could be no result. If one were to try to produce a rectangle by multiplying base (b) times height (h), this figure could not be made, for it would be a product of incompatible numbers. Similarly, a formula like $1 = x^2/a^2$ & y^2/b^2 would be impossible, for it would imply that $1 \sim x^2/a^2 = y^2/b^2$, which would mean that although x is a real number and y is a veritable number, or *vice versa*, the two expressions could be equilibrated—an impossibility. Not only are the negatives of the two number systems incompatible, but even the positives. If a particular positive number— say +9—were real, it could as easily have been the square of +3, as of -3. But, in the veritable system, (-3) x (-3) = (-9). The two kinds of numbers are not interchangeable in any way.

Veritable numbers can be changed into absolute numbers and so can real numbers, but they cannot be set into a proportion like this: v/A :: v/R. Just as it is only when one is working within the real system that absolute numbers can be converted into real positive numbers or *vice versa*; similarly, it is only within the veritable number system that veritable numbers can be converted into absolute numbers or the reverse. Yet, even there, a positive veritable number that had been converted into an absolute number could not be transformed into a negative veritable number. There is no interchangeability there either. Identity is preserved throughout. The absolute works as a positive in the real system and as either positive or negative in the veritable system, depending upon which one is being used. To summarize: while there is no incompatibility if a single point has veritable and real components, no finite construction whatsoever is possible between them.

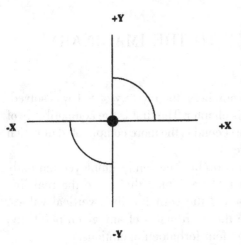

Veritable positive and veritable negative numbers cannot be used together in rectangular multiplication, division, addition, subtraction, or extraction of roots. Consider the drawing on the left. Both coordinates are veritable. The figures on the 1st and 3rd quadrants are quarter circles. This is possible because the horizontal and vertical coordinates of the first quadrant are both positive, and those of the third quadrant are both negative. The horizontal and vertical coordinates in the 2nd and 4th quadrants are mixed, $\mp$ and $\pm$, respectively.[166] Given standard Cartesian formulae, nothing can be drawn there. The coordinates exist, but they cannot be connected. They are not even potentially lines, for no part of them can approximate a line. Yet, the lines in the two possible quadrants begin and end at the exact same points where they would have, had a full circle been capable of existence.

In the last chapter, the real and veritable number lines were represented as parallel. Since parallel lines are by definition equidistant at corresponding points, there is no possibility of interaction there. This would be the case with straight parallel lines, concentric circles, and any two or more wavy lines of the same curvature and size lined up in the appropriate orientation.

Is there any other orientation which is theoretically possible?

Suppose that the two lines, one veritable and the other real, were extended toward each other in opposing directions. Would they be compatible? They would indeed, since when they met, they would not interact. They would merely touch each other. Since no point could be placed between them, there would be no possibility of any cross breed. Furthermore, since they were being extended from opposite directions, they were really two lines. And since the beginning point of each, or zero, is on the opposite end, the line would be incapable of further lengthening in the direction pointing toward the other line.

Suppose, then, that instead of crossing each other like a pair of Cartesian axes determining the area between them, the two lines were simply perpendicular ($\perp$), with one line resting upon the other. Would they be compatible? They would, indeed. A veritable and a real line which merely touched one another would be compatible, but they could never be used to construct a figure on the basis of a common coordinate system. Whichever axis was vertical, the 1st quadrant would be impossible because the positives would not be compatible; the 2nd, because of doubly mixed signs; the 3rd, because the two negatives would be incompatible, and the 4th, for the same reason as the 2nd.

Suppose that the real and the veritable number lines had not been constructed using Cartesian axes, but the two lines did not merely touch, but intersected at right angles, and the point of intersection was labeled "0". Is this possible?

This requires some discussion. Zero in the real system is not used the same way as it is in the veritable system. The zero on the real line is used in a relative way, in that negative values are considered to be less than positive values, although the magnitudes are the same. For instance, the zero on the Fahrenheit scale does not signify the extinction of all heat intensity, but only less than a positive temperature and more than a negative Fahrenheit temperature. The zero on the veritable line is not relative in that respect, for neither direction is greater than the other. In that number system, a negative number of -100 is in no sense smaller than a magnitude of +100. Zero is not used in the same way in the two systems.[167]

In the case under discussion, however, it would not make any difference if an operator would use the zero point differently; the two incompatible lines cannot interact anyway.

Suppose that the two lines of opposing natures cross each other at some angle greater than 0 but less than $\pi/2$. Consider this possibility. A point in the space between them would have been determined by two coordinates, one of which would be veritable and the other real. As stated above, it would be logical for the same point to have these determinations. But a complete figure using all four quadrants would be impossible—for the reason given earlier.

Calculations would, however, be possible if they concerned either the veritable line by itself or the real line by itself. Although the two number lines were incompatible, since they shared no finite extent, this would not affect either. If, for instance, the point on the veritable line were +2 and that intersecting point on the real line were +2, this coincidence would not change any arithmetical operation which pertained to the one but not to the other. Squaring a real positive two would not affect in any way the square of the veritable positive two. It is the particular numbers of the two systems that cannot be combined; and those magnitudes are not set by a single point. To give another example, the finite distance between the real (-3) point and its 0 establishes its nature, not any relationship between that point and, in this instance, the quite different direction of the veritable magnitude -3, as measured from its 0.

It is impossible to construct any figure using pure real and veritable lines together as coordinates, but it has been determined that one or the other, or both, can be made to approach each other by any angle. The same arguments would apply to a minimum angle of one infinitesimal.

Suppose next that the veritable and real lines use the same zero. Suppose further that the magnitudes are the same in the two measures, but the veritable and real signs retain their identities. Is that possible?

It certainly is possible. Just as a single point on a circle can be labeled as both "0" and "2π", so *it is possible that the same point on the same line can have distinctly different indications, providing that there is no commerce between them.*[168]

That is how it is with an imaginary line. Real and veritable numbers cannot be added, subtracted, multiplied, or divided together rectangularly; neither can they be mixed in finding roots or establishing ratios. Yet, the imaginary axis is a combination of the real and veritable numbers.

How can these two incompatible types of numbers be combined on a single number line without contradiction? How is it possible for the imaginary system to exist when its two components, real values and veritable values, are ineradicably different?

We know that imaginary lines exist. It remains to find how their construction is possible.

1st The imaginary line is not a real line put next to a veritable line. If that were the case, veritable and real values could not have a single expression. Therefore, there must be only one line, not two parallel lines touching each other throughout.

2nd The two lines cannot be coincident. This is impossible; it would contradict the very nature of an infinitesimal, no two of which can exist at the same place, since they are in two separate and distinct places.

3rd The veritable and real determinations must be on the exact same line, simultaneously. The same point on the line can have determinations in both systems without contamination.

4th The square root of negative one has the same magnitude as +1 and -1 in the veritable number system. It also has the same magnitude as ± 1 in the real system. Both are based on the absolute number system. Thus, the two incompatible systems have the same magnitude.

5th Although √-1 has the same magnitude as ONE in all three systems, it is not identical to any of them. It is not even identical to -1 in the veritable system, even though they have the same magnitude. It is still only its square root; it is a function of the other.

6th The veritable portion of the imaginary line is not a continuous line of ascending values from 0 to + 1 on the positive to progressively higher values and the reverse on the negative. It is rather a stack of the $\sqrt{-1}$ on top of each other. There is increase, but it is simply a count of the number of i's, or their fractional and non-rational parts placed on top of each other. *The count is in real numbers.*

This peculiarity of construction leads to the result that one type of multiplication is impossible on the imaginary line. Rectangular and rotational multiplication have already been discussed. There is a third type that can be performed on an ordinary line, but not on an imaginary line. This third type is linear multiplication. If a line is multiplied by a magnitude equal to itself, it is also squared. Linear squares are not direct areas, nor is an area indirectly implied in their generation. For instance, $(+2)^2 = +4$, where both +2 and + 4 are lengths on the same line. It is possible, then, to produce a linear square on the real axis. It would also be possible to produce a linear square on a pure veritable axis. There, a magnitude +2 multiplied by itself would yield a longer line of + 4; also, of course, $(-2)^2 = -4$. (As was dramatized in that late 19th century work, *Flatland*, the product of two lengths need not be a rectangle.)

Because the imaginary line is a stack of $(\sqrt{-1})$'s, it is impossible to produce a linear square of i, directly. Recall Figure ii in the previous chapter where the real number system is represented as a horizontal line; the range from -3 to +3 is shown. Note that along this axis, it is possible to express the square of any real number, such as 2 x 2 or $2^2 = 4$; the same would be the case with any number of the veritable axis; $-2^2 = -4$. Yet, along the imaginary axis, i^2 cannot exist. One can give i any real factor or coefficient one wishes, 1/5, $\sqrt{5}$, 5, -5,000, etc. But one cannot express the linear square of $+i$ on the imaginary axis, since its real product, -1, is not and cannot be on that axis. This is because of the peculiar nature of the imaginary numbers. If such a square were attempted, it would be -1. But -1 is a veritable number that cannot exist on that axis either. Each unit of i can be divided into rational and irrational fractions, but the unit itself cannot be squared in veritable terms along its own axis. It is a stack.[169]

7th The i's are stacked one on top of another. At some point, the stack is divided in two parts and separated by an infinitesimal, which is labeled by a real "0." This is very important, because it destroys the sameness of the line without rupturing continuity on either side. On the plus side of the imaginary line are the same negative square roots as those stacked on the minus side. This prevents the possibility of i being handled as a veritable number; the place for the positive veritable on the other side of the zero has been pre-empted. The position of zero also prevents the possibility of their being handled as absolute numbers, since their enumeration requires signs, both positive and negative.

8th The $(\sqrt{-1})$'s on the now negative side of zero do not change their inner veritable sign or their magnitude, but are automatically pointed in the opposite direction to the real positive. Smaller divisions are simultaneously reoriented.

9th Each square root of negative one magnitude on the imaginary axis is, simultaneously, labeled with a real number coefficient, positive or negative. This extends as high as the imaginary axis is brought up, or as low as it is brought down. Each fractional or irrational part of any of these i-magnitudes is potentially labeled in real numbers. Although the real and veritable portions have distinct identities, they are organized to operate together as a single unit of a coalescent nature.

Every single point on the imaginary number line is both real and veritable. Each point is specified by its real component. But neither the particular i on the stack is identified, nor its rational or irrational designation on the i is given. To illustrate, $+2i$ identifies the number as the second level of i's in real terms, but not in veritable terms; similarly, $-\sqrt{5}\ i$ identifies the level of i's and specifies which point it is on that level in real terms—*the real square root of five* imaginary units down on the negative axis. Since the veritable gives an incomplete identification of a point, the real is required to supplement it. As X is the only real axis, the i part shows what axis it is on, the Y axis. Because the magnitudes of the real and the veritable units are the same, there is no incongruity with respect to size. The partial contribution of each to the identification of any point on the line helps unify it. This union is not potential, but actual.

10th This union is sealed by the fact that the re-arrangement of the stacked i's with the infinitesimal interpreted as a zero in its midst alters the very nature of the type of multiplication possible in the stack. When a pair of real number lines are crossed in a standard Cartesian coordinate system, two types of multiplication are possible. The first is rectangular multiplication, where a value on the X-coordinate is multiplied against a value on the Y-coordinate, producing a product. The other is the just discussed linear multiplication. Neither of these is possible with the real-imaginary cross.

A glance at the first figure in the previous chapter readily shows that the result of this rotational squaring is the real value of -1 on a line at right angles to the imaginary axis. The lengths of the units in the two axes are the same. If one examines the course of the wheel as it turns from $+i$ to -1, the arc which it describes is inside and therefore less than a unit square formed by $+i$ and -1. Because of its incompatible elements, rectangular multiplication is not possible, so an arc is described instead. It can only rotate.

11th All this takes place within the unit circle as described in the last chapter. The generation of the rotational squaring across two neighboring quadrants from a real +1 to imaginary i to i^2—a real -1—can be measured by the Eulerian formula. In the rotational square, i.e. at -1, all of the veritable element is fully expunged and made equal to 0. In other words, the rotational multiplication begins with a real value +1 and an imaginary value of 0; then the imaginary value increases and the real value declines. At midpoint,+i, the real element is zero, and the imaginary value is at its full value of +$\sqrt{-1}$. After that, the imaginary value declines and the negative real value gets greater. Finally, the rotational product is a real -1 with a veritable value of zero.[170]

12th Yet, i has been "squared." The area formed by the arc in the first quadrant between +i and +1 has the same area as the one just described in the second quadrant; the angle formed by the two adjacent arcs is twice $\pi/2$. This, we have designated as the "rotational squaring" to distinguish it from the two plainer varieties. Sweeping across is the radius vector, which being stated in absolute numbers, can work equally well with both. The veritable and the real alternatively rise from zero and then fall again. The whole resembles a mechanical watch with the contrary parts balancing and reinforcing each other.

This is how the imaginary line is possible. The mathematicians who came up with the imaginary system did not proceed this way. They did not know of the existence of veritable numbers. Instead, they worked from the back with what they knew in order to handle an inescapable contradiction. What they did with it was something wonderful to behold.

What was to them still a mystery with the interpretable issuing out the uninterpretable is apparent to those of us who understand veritable numbers. The stacks of i and the real numbers which designate its points have the same origin or "0."

To repeat what was said earlier: in the case of the veritable and real lines that intersect at some angle where the point of intersection is designated in both veritable and real symbols, there is no contradiction—no contradiction, because there is no standard arithmetical operation using both of them, simultaneously, i.e., they are not added together, subtracted together, etc. To explain further: the point by itself has no value, but must refer to magnitudes which lie between it and the separate origins of the two inclining lines; each number line is at an angle with respect to the other. For that reason, the two evaluations are mutually exclusive alternatives. The point cannot hold a fusion of both evaluations.

In the case of an imaginary line, the same magnitudes are marked in both veritable and real evaluations. There is no contradiction. As long as they are not confused in some way, the same point can receive both designations. For instance, a point on the imaginary axis can receive both a real designation of "+1" and a veritable designation of $\sqrt{-1}$; the two can be identified together as +1 $\sqrt{-1}$ or, more familiarly as +i. Since the imaginary line is only a repetition of the same veritable magnitude, unit after unit along the number line, this allows for computation from the real aspect. No contradiction exists when the real magnitude is equal in length to that of the veritable. With the repeated veritable quantity serving as the background and the real designating the numerical and positional values on it, the two are united into a single new unit, the "imaginary."

Not only are limited computations possible along the imaginary number line, but through the Eulerian and De Moivre formulas, computations of the juncture of real and imaginary lines can be made in a unit circle. On that circle, the angle of a radius vector for any successive roots may be found directly. $+\sqrt{i}$ for instance, can be found directly at $\pi/4$ and $+\sqrt[4]{i}$ precisely at $\pi/8$. Indeed, every point along the periphery of the unit circle will have both a real and an imaginary component. This is also the case with any point of smaller radius.

Earlier discovery of the veritable numbers as the basis for the imaginary numbers was obscured by the very fact that $+i^2 = -1$. Since they did not know what a veritable number was, it was natural for them to conclude that the negative one in this square root was the same as the minus one that they already knew about. Yet, even so, they knew that there was something peculiar about it; and so the symbol "i" was brought in. Unfortunately, the symbol itself brought obfuscation. The fact that it could be used as a constant in the above mentioned formulas prevented them from realizing that it was the square root of another kind of number.

Imaginary numbers were discovered in the solution of third degree equations; They were not posited, but discovered. The veritable number was the answer to the question, what kind of number would have to exist in order for there to be a square root of negative one? Both induction and deduction played a part in each discovery. But the act of discovery itself was crucial.

CHAPTER XV

ON APPLYING VERITABLE NUMBERS

In what way does the distinction between these kinds of numbers bear on a study of the infinitesimal? It is this: direction is a fundamental attribute of a line. It is not the same thing as a line, but it points from its origin outward. Once zero is established, the rest is possible. The three ways of handling direction result in the absolute, the real, and the more recently discovered, veritable.

In this chapter, veritable numbers will be discussed in relation to the following: the differentiability of geometric figures, negative exponents, the negative numbers themselves, ratios, and physical science.

Consider first, *differentiability*. Let us examine the well-known equation, $y = f(x) = \sqrt{x}$. According to the standard mathematicians, Richard Courant and Fritz John, this equation "is defined and

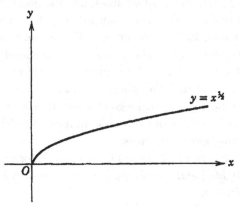

continuous for $x \geq 0$, is also not differentiable at the point $x = 0$. Since y is not defined for negative values of x, we consider the right-hand derivative only. The equation $[f(h) - f(0)]/h = 1/\sqrt{h}$ shows that this derivative is infinite; the curve touches the y-axis at the origin."[171]

Such a figure has a derivative in the veritable system. For negative values, the equation is defined as $y = f(x) = \sqrt{-x}$. It also has a left-hand derivative at $X = 0$. This shows that the veritable system has some applications not present in the other one.

Let us next consider how the two systems handle *negative exponents*. In the real number system, there is a difficulty of representing exponents for negative numbers. Some like Euler are of the opinion that it is an imaginary number, i.e., (-2^{-x}) is complex.[172] Others, that such an operation cannot be carried out[173], and others simply leave out all discussion.[174] The majority opinion is with Euler. This leads to the subject of complex analysis, which is beyond our purview.(The author may take this up in an another work.)

In the veritable system, this is straight-forward. An expression like $-a^{-x}$, where $-x$ is a negative exponent, is as easy as dealing with positive numbers with positive exponents. The imagination of the theorist is not taxed with the thought that there might be some inherent resistance in reality to a negative line of force.

Let us now pass directly from exponents to a general consideration about *negative numbers in general*. The 17th century mathematician, Antoine Arnauld, rejected negative numbers. His reasoning was as follows, according to the division rule used for negatives, -1 is to +1 as +1 is to -1, or -1/+1 :: +1/-1, which means that a lesser number is to a greater as a greater is to a lesser.[175]

The veritable system makes better sense than the real when dealing with ratios. According to a 19th century source, "In arithmetic it is the measure of the relation of one number to another of the same kind, expressed by their quotient."[176] This is the same as Euclid's definition.[177]

Consider the real ratio, -3 : -4. In the real system, this is 3/4, which is the same as +3 : +4. This measure of these numbers of same kind is accurate only with respect to magnitude. In accordance with its rules, the sign disappears. Yet, in the real system, -4 is actually smaller than -3. The reason, as stated in Chapter XIII, is that "negative numbers are considered to represent smaller quantities as well as opposite directions; this can be seen from the fact that in order to move from the negative to the positive one must add; conversely, to move from the positive to the negative side, one must subtract, i.e., the quantity must be reduced. In the veritable number system, by contrast, negative and positive signify direction only, never size. In that system $-a > 0$. Any number added to negative one in the veritable system can only be one of greater negativity. The total veritable operation is like the central axis of a double cone extending outward; nothing from one side can interact with the other."

The real number handling of the ratio does not make good sense. If -4 is less than -3, the transformation to 3/4 is not right. In the veritable system, -3 : -4 is expressed in the same terms as the numbers themselves.

In the real system, -3 : +4 = -3/4. How is it that the positive denominator was removed? The difference of sign is slighted and the comparison consists only in an expression of their relative magnitudes.

The veritable system is not like that. Although a direct comparison of a ratio with mixed signs is not possible, they can be converted into absolute numbers. With the veritable system, however, there is no distortion on the finite level. A ratio like -3: +4 becomes |3:7|. This is because the actual difference between -3 and +4 when converted into absolute numbers is three units on the negative side plus four units on the positive side. Since there is no reflection in this system, this actual finite difference is not denied in the process of conversion. Note that the ratio sign is preserved. With mixed signs, the veritable is a better measure of the relationship than its real rival.

Real ratios tend always to be expressed as division. When the signs are the same, they are quickly reduced to lowest terms; +16 : +32 becomes ½. There is no distortion, but the consideration of the actual magnitudes involved is dropped in lieu of a relative comparison. The major exception is when dealing with irrational numbers. In $\sqrt{2}/2$, for instance, since the numerator is incommensurate with the denominator, it is sometimes left as a ratio; more often, however, a digital approximation is substituted.

This may be why it is that ratios have been all but replaced by division in modern science and mathematics. With the introduction of the veritable, this could change. Veritable ratios can be expressed in lowest terms, but always by preserving their nature.

Logically, one direction cannot be turned into another. They are what they are, orientations in space. When the real number-boomerang turns, it is no longer going in the same direction. Where positive and negative actually refer to opposite directions, the veritable is the right choice. The fallacy noted by Arnauld does not exist there. This is because negative and positive numbers are equal in metaphysical significance.

Let us next turn to *physical science*, especially bodies in motion. Consider first the subject of translation. Suppose that opposite forces should exert themselves against a block on a flat, frictionless surface. Let the impulse to velocity from the left be +6 units per minute and the one from the right be -3 units/minute. The net velocity would be +3 units/minute. This could be obtained in either system. In the veritable system, this would have to be obtained by first converting to absolute units (which are directionless), obtaining the difference, and then identifying by how much the veritable positive was greater. In the real system, this can be done more promptly, because of the attribute of reflectivity. If one were looking at it from this perspective only, one might think that problems involving linear forces are handled more quickly with real numbers. But, as we shall see, there are some important exceptions.

Suppose this time that the force from the right were $\sqrt{-5}$ poundals. In the veritable system, this would mean an irrational force in the negative direction. This is impossible to state in the real system. In that system, a force like that could not come from the right.

A distinguished physicist pointed out to me that a person using the real system could, indeed, employ the position suggested by (-2.3461) as a rough form of $-(\sqrt{+5})$, i.e, by subtracting the square root of +5 from 0. This is true as far as it goes, but it does not change the situation: the square of the real (-2.3461) is always and forever more positive in that system. It may parallel the same number in the veritable system, but its square can only approximate +5, never -5. More important, $-(\sqrt{+5}) \neq \sqrt{-5}$ in the real system. The comparable quantity, $\sim(\sqrt{+5})$, exists in the veritable system; it simply means that the positive square root of five is to be subtracted from a veritable number equal or greater to it, and the difference must be zero or a positive. This important difference cannot be understood in pure real number terms. Moreover, the problem for the real theorist is a negative square root force, not a mere exercise in abstract subtraction.

According to real number system, the only meaning a negative square root could have in a situation like this is that it have a motion at 90°, ±, from the path. Negative square roots—or rather, a modified form of them, the *imaginaries*—have to be represented at right angles to the real line. This is because they contradict the real numbers with respect to multiplication and division and, by implication, addition, subtraction, ratio, and extraction of roots as well. Attempting to incorporate them into the real system would be like trying to pass off an apple as a kind of orange on the grounds that both are roundish, seed-bearing fruits with some sort of a reddish skin. Accordingly, imaginary numbers are commonly listed as an adjunct to the real number system.

But this sort of a representation beclouds the truth that the actual force is not operating at right angles from the point of application but is in line with the positive force, although opposing it from the other side. The veritable representation expresses the actual facts, while the real system founders, even when it has help from the imaginary numbers, which are hybrids of the two systems.

And finally, $\sqrt{-5}$ is not the same thing as $5i$. The positive imaginary axis is a stack of $\sqrt{-1}$'s. It does not have a negative five inside a square root symbol. Properly speaking, it does not exist in that system.[178]

Let us turn from translation to rotation. It appears that either number system could handle a counter-clockwise rotation of $\sqrt{+2}$ units /minute[179] and a clockwise one of -1 units /minute opposing each on the same axis; the former one would have proceeded more than 1.414 times as far in a minute's time, and the net effect would be roughly +.4141. The question is, which one is most appropriate?

Consider the rotation of a wheel on a single axis, such as might be found on a mechanical safe. Let us call the counter-clockwise "+" and the clockwise, "-." From the right, a turn is given of +6 units along the dial; and from the left, one of -3 units. Any additional turn given to the negative—for instance -2 units—would add just so much of a turn to it in this direction—in this case, to -5 units. A contrary turn of +2 units would produce a positive turn of +8 units/minute. Or, to put it differently, an augment in the positive direction increases the counterclockwise count; any augment to the negative means that much more of a turn clockwise. Which numerical system better describes this kind of phenomena?

Consider more closely this case of the dial in which, once again, a counter-clockwise motion is positive and a clockwise motion is negative. Let us look at it first in terms of the real system. Suppose that two clockwise clicks are made, resulting in (-2), followed by three positive movements (+3). Here, (-2) + (+3) = +1. In that vein, someone might argue that the past actions that had produced the two negative unit turns were canceled. But this could not be meant literally; one could not successfully argue that the two negative turns were reversed, for this did not physically happen. They had already been completed. They were no more. What happened, the exponent of the real system would say, is that they had the net effect of one counterclockwise unit. But was this truly the effect? If it were a combination lock, producing the single counterclockwise click would not open the lock. In such a case, there is no cancellation. Any previous movement is essential to the final result. If—to change the illustration—a reciprocating engine were shutdown on a downstroke, would this give one a better picture of its performance than the total upstrokes and total downstrokes? Obviously not.

But devotees of the real system could argue that while it is true that the net is frequently not the most important part of the operation, the real system is not contradictory, since one can think of the above as two negative clicks followed by three positive clicks, resulting in a net positive. But what would be the utility of thinking like that?

Let us attempt it in the veritable number system. Any turn clockwise is negative; any turn counterclockwise is positive. There is no need for a net. Clearly, the veritable solution is the more appropriate in the case of a combination lock.

Consider not a lock, but a wheel free to turn in either direction on its axis. Suppose a rotation of 6 units counterclockwise (+) followed by a rotation of 3 units clockwise (-). If the real system is employed, this would be considered as either (a) a rotation of six positive units followed by a subtraction of three positive units, leaving a net three positive units; or (b) a rotation of six positive units, to which are added a rotation of three negative units. If (a), then both clockwise and counter-clockwise are regarded as positive. If (b), then counterclockwise is + and clockwise is -. So far, this is self-consistent. It amounts to arguing that $(+6) - (+3) = +3 = (+6) + (-3) = +3$.

What about the subtraction of a negative? Can one subtract three clockwise units from six counterclockwise units? Can one, for instance, do $(+6) - (-3) = +9$? Only if the clockwise three could reverse itself ; and to accomplish that, these clockwise units would have to be their own opposites, which is impossible.

Therefore, when it comes to rotation, $(+6) - (-3) = +9$ is not the same as $(+6) + (+3) = +9$. The two are not simply two ways of stating the same thing. Therefore, the real system is confusing when applied to free rotations, confusing because it cannot be applied consistently.

The veritable system is more appropriate. In that system, one can only add or subtract units of the same sign. An addition to the negative makes it more negative, i.e., clockwise. In the case of the freely turning wheel, subtracting the negative rotation does *not* have the same effect as adding to the positive. The negative does not move out of the plane and reverse itself. Subtracting the clockwise is not the same as adding to the counterclockwise; to put it differently, & (+) is never equated with ~ (-) in the veritable system. The positive and negative sides of zero can never mix. The veritable system applies consistently and thoroughly to the phenomenon of that wheel. The real system is not as thorough.

To understand why it is like that, let us turn to the different way that the two numerical systems handle directions. In the real system, directional signs can sometimes be used interchangeably with addition and subtraction signs. The reason, of course, is that in the real system, negative numbers are lower than real numbers, despite their equality with respect to magnitude. In fact, in real formulations, the plus sign is frequently skipped in the case of an addition of a negative.

With the veritable system, one cannot add to or subtract a negative number from a positive number; neither addition nor subtraction may be performed with numbers of the opposite kind. With the veritable, one can only add negative numbers to negative numbers and positive numbers to positive numbers; the same with subtraction, multiplication, and division.

What about the sums of magnitudes that go across the signs—say from -2 to +3? Is the person working in the veritable system denied such knowledge? Not at all. To register such a change, one must first convert into absolute numbers and then give their sum—in this instance $|5|$. One can also register the equivalent of a net. Since a positive veritable unit and a negative veritable unit have the same underlying magnitude, there is no contradiction is stating which magnitude, the + or the - , was greater. One is not precluded from saying that the magnitude underlying positive three is greater than that of negative two by an absolute magnitude of one, i.e, $|1|$. Simple calculations like that can be performed in one's head.

In what respect is the veritable system superior? In this system, there is no built-in bias in favor of either direction, as there is with the real system. This built-in bias can also be found in the way the real system handles absolute numbers: such numbers are treated as if they were positive. In the veritable system, by contrast, absolute numbers are converted into negative when one is working with negative numbers and into positive when one is working with positive. Quoting again from Chapter XIII: "There being no predisposition in favor of the positive sign, an absolute number used in the same operation with a negative number would be treated as a another negative and one used with a positive number would be positive. Thus $(-11) \sim |7| = -4$ and $(+11) \sim |7| = +4$ in the veritable system." It is this absence of bias at the very foundation of calculation that should be weighed by the scientist.

In the real system, addition of negative numbers always means a lessening, even when one adds a negative number to a negative number. For example, (-3) + (-2) or (-5) is less than (-3) - (-2) or (-1), since the latter is closer to the positive numbers; in that system, zero is really larger than any negative number. In the veritable system, by contrast, zero is neutral, signifying the absence of the negative as much as the positive. Any number of either sign is greater than zero. It follows that the veritable system is better at showing forces of contrary directions than is the real system.

What is the significance for physical science in general? Let us consider the real system first.

The real system is appropriate for any field of study in which the part to be labeled "negative" in some sense represents a diminution of an attribute(s) signified by the positive. For instance, in the Fahrenheit and Celsius temperature scales, each lower degree means a reduction in the average intensity of molecular activity. A sub-zero reading in Celsius, for instance, means that there is less of that heat intensity than at the freezing point of water under standard conditions. Typically, although not universally, the "0" in such cases is arbitrary; it would have been possible to make something else stand for zero on the Fahrenheit or Celsius scales. The Kelvin scale, however, (although it incorporates Celsius-sized units) does not use the real system. It uses the absolute number system out of which both the veritable and real systems emerged. It is the same as the positive side of the veritable system.

The real system is appropriate when the zero is not totally neutral. This is the case with money. Zero money means none whatsoever. Negative money, or debt, is of dubious value in comparison with the real thing. To repeat, Euler once justified equating the subtraction of a negative number with the adding of a positive number with the example that "to cancel a debt is the same as giving a gift."[180]

The real system is appropriate when using coordinate systems with four quadrants. The third and fourth quadrants can be represented there, which is not the case with the veritable system. This has already been shown to be the case with the Cartesian system. The use of the veritable system with polar coordinates is also attenuated, although not to the same extent.

It would be a great mistake to use the veritable system in such operations, since, in doing do, one would be presuming a metaphysical equality between positive and negative where it is not appropriate.

The veritable system is of greatest value in dealing with natural phenomena in which the negative is not a diminution of the positive, but is of equal power. In the case of the combination lock or a wheel on an axis, it is obvious that the clockwise rotation is in no sense deficient in some quality possessed by the counter-clockwise. The point at which a wheel reverses direction is zero, and an intermingling of opposing directions is impossible. With the discovery of veritable numbers, there is no further need for those who know both to continue with the inappropriate use of real numbers.

This may also be the case with magnetism, where both poles exist. There, reflectivity does not pertain. For example, if one starts with a magnet in which the south pole of has a strength of -3 and then, somehow, its strength is decreased by -2 units, this cannot be described as -3 -(-2) = -3 +2, even though the result of (-1) is on both sides of the equation. The negative pole exists in its own right, so to speak, and is not a mere diminution of the positive. Therefore, the second part of the equation (-3 +2) does not belong. Prior to the discovery of the veritable number system, this difference was overlooked. The mind simply passed over the fact that the real system did not properly apply there and employed what belonged to the undiscovered veritable system.

Another instance is Newton's third law of motion in which the two forces are equal and opposite. The opposing but parallel forces entirely fit into the situation where the veritable number system works best. The reactive force does not intermix with the initiatory force in a reflective manner. Reflectivity, it will be recalled, is this property that real numbers have of doubling back or reversing the direction of their sign. Prior to the discovery of the veritable number system, the exclusive use of real numbers was justified because of ignorance. It may be that the reason why so many scientists have trouble with Newton's third law is that it is difficult to comprehend the idea of equal and opposite forces using a number system in which the reactive side would have to be less.

The veritable system can represent figures fully in coordinate systems based upon two sections, one positive and the other negative. This writer also thinks that it can be applied to the vector, resulting in something midway between Sir William Hamilton's quaternions and Josiah Gibbs' system. Since every veritable line is the equivalent of a vector, this fact alone may prove of some assistance in physical science. There may also be some applications with ordinal numbers.

Furthermore, giving the negative the same metaphysical status as the positive and opening up mathematical science to the direct calculation of roots may do much to bring physical science into a place which it is hesitant to enter. Opening up what was formerly partly occluded may bring about a hitherto undreamed of understanding of nature.

And more, since in the real system, the negative is only a diminution of the positive, using it exclusively may lead the mind to unnecessarily conclude that nature reduces to more of a Cosmic One than it is.

The discovery of the veritable system is recent. A more detailed study is being planned. This should be enough to indicate some ways in which direction, a key attribute of lines can be applied when used in connection with numbers.

CHAPTER XVI

THE TANGENT

A tangent is a straight line which touches the surface of a curve at only one point. This property makes the tangent of almost inestimable importance.

In the 7th chapter, an ambiguous figure was introduced, which was interpreted as either a continuous line with loops on opposite sides or two circles touching a common line. If the line was a tangent, then it was separate from the two circles; if the line did more than merely touch the two circles, then the latter were loops. This shows that it makes a difference whether the line in question is a tangent or not.[181]

There are some who would question this definition of a tangent. In Rosenblatt and Bell's *Mathematical Analysis for Modeling,* for instance, the tangent is shown skewering the line, instead of lying on it.[182] A modern dictionary, *The Universal Encyclopedia Of Mathematics,* supports these authors: "A tangent is a straight line which has a point in common (point of contact) with a given curve (of arbitrary form) and at this point has the same slope as the curve."[183] The phrase, "point of contact," is ambiguous, but the rest of the definition makes it clear that what is intended is a point of connection. To the left is a drawing showing this idea at C.

Yet, the Latin, *tangere,* means to touch.[184] Although Euclid does not give this curve a name, he certainly identifies it as existing outside but touching a curve. (*Book III, Proposition 16.*)[185] The equivalent words in Archimedes, ʹεφαπτομένη and

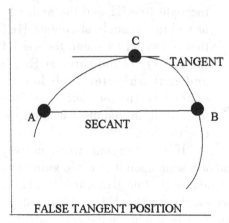

C

TANGENT

A

SECANT

B

FALSE TANGENT POSITION

ʿεπιψαύουσα (ephaptomene and epitsauousa), mean to touch a curve or surface.[186] Sir Isaac Newton did not discuss this aspect of the tangent is his *Principia*.[187]

Which is it? Does the tangent lie on the curve or does it skewer it? This makes a difference when analyzed in terms of infinitesimals.

A tangent is not simply a line which meets a curve at one point without meeting any other points of that curve—as that *Encyclopedia* asserts. In the drawing above, more than one such line can be drawn through point C. In any open curve, at least one second line can be drawn through the opening. In fact, multitudinous lines can be drawn swinging from a point in a curve. But the tangent must be *unique*.

The uniqueness of the geometrical tangent can be shown from the following consideration. Suppose first a straight line. Then a curve touching it. No other straight line can be drawn between the two neighboring points without putting the original straight line out of the way and to the side.

In his foundational study of the calculus from a geometrical standpoint, *A Treatise On Fluxions*, calculus pioneer Colin Maclaurin showed that the achievement of Sir Isaac Newton was consistent with the work of Archimedes. Although Maclaurin did not accept the reality of the infinitesimal—in fact, his treatise was written in opposition to its idea—he *started out* by affirming that the tangent must touch. "As a rightline is the tangent of a circle, when it

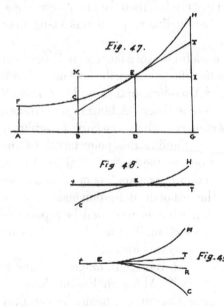

touches the circle so closely that no right line can be drawn through the point of contact betwixt it and the arch, or within the angle of contact formed by them, so, in general, when a right line ET (*fig. 47 modified*) touches any arch of a curve, as EH in E, in such a manner that no right line can be drawn through E betwixt the right line ET and the arch EH, or within the angle of contact HET that is formed by them, then is ET the *Tangent* of the curve at E . . . . and that when the arch has its concavity turned one way, the tangent at any point of it is on the convex side."[188]

If the tangent were merely intersecting the curve at point E, instead of resting upon it, it could swing into more than one position; for example, it could take the line MEI without touching the curve a second time. It could also occupy the position E D and then extend itself on the other side of E. The tangent must be unique to the curve which it touches.

The importance of this fact follows from the following consideration: It is reported that Leibniz tried to approach the idea of the calculus through the secant. His idea can be illustrated by referring to the first drawing above. Let a secant be drawn through A on a curve to some arbitrary point B. Then, move this secant parallel to itself toward C along the curve. At point C on the curve, the secant is supposed to suddenly transmogrify into a tangent. As Friedrich Waismann put it: "Up to the time of Leibniz there existed in mathematics a deep chasm between secant and tangent. For example, propositions valid for the secant of a circle are totally different from those valid for the tangents of a circle, and it would have occurred to no geometer to establish propositions common to these two kinds of lines. It seems that Leibniz, through his philosophical ideas regarding a continuous connection of all things, being his *loi de continuité*, was led to the view that tangents could be adopted among secants as limiting cases. At any rate, this kind of vision indicated the hours of birth of differential calculus."[189]

This is fallacious. Between the geometrical tangent of a point on a line and any shrinking secant, there must be the line itself. The interval between the secant and the line gets smaller and smaller. Eventually, when there is no finite distance between it and the curve, the interval is destroyed. Then there is nothing but mere locations between the collapse of the interval and the curve; and finally, the point on the curve. Then, on the other side of the curve is the tangent. Because of the infinitesimal which must lie between them, the secant line can never be identical with the tangent line.

This is very strange, since Leibniz is thought by many, including Abraham Robinson, to have been a champion of the infinitesimal.

But let us continue. Those opposed to the idea of a touching tangent line might bring up the case of the point of inflection, since the tangent line appears to go through the point of tangency.

Consider Colin Maclaurin's ambiguous discussion: "The right line TE (fig. 48) being continued to t, if Et is the tangent of the arch EC the continuation of HE, then the arch HEC has a continued-curvature at E. When the arches EH and EC are on different sides for the tangent Tet, the point E is called a point of contrary flexion."[190]

In ordinary construction, the line is drawn right through it. Many would argue from this that this provides an exception; that E, a point on the line, is also a point on the curve.

Against this, the classicist could argue: The tangent line actually goes from point T till it touches E; there is a second line in the opposite direction from t touching E from the left. Either one of these lines measures E's derivative to be 0, since at that point, the upper flexion begins to bend from the point above it through H and the lower flexion, from the point below it through C.

Actually, the other interpretation often has the true position. Consider the calculus of the derivative. Its tangent is primarily trigonometric, rather than geometrical. $\triangle y/\triangle x$ is *that* kind of tangent.

Crucial to it is the angle between the hypotenuse and the base ($\triangle$x) in a triangle with $\triangle$y as the vertical side. This is the differential triangle. Within it can be found a non-trigonometric tangent. This is because the hypotenuse acts as a tangent-line. It joins a point on the curve to the top of $\triangle$y and relates both to the beginning of $\triangle$x. Let us call it the *hypotenuse-tangent*.

It will be recalled that in the minimum triangle, the opening is formed by its sides, and that these sides are external to that opening. Any angle of any size can be put together that way.

But in a standard Cartesian set-up, this will not work. There, the sides of the angle do not merely frame it; they are part and parcel of it. For example, the vertical line in the right angle of any quadrant belongs to the angle quite as much as do its interior parts. A framed angle serving as a hypotenuse-tangent in a Cartesian configuration would be one infinitesimal too large. That is because it would be lying *outside* the curve. The angle would be the tiniest bit too large.

The only solution is that the hypotenuse-tangent not rest upon the curve, but actually *go through it*. This is possible, even though it contradicts the geometrical tangent. It is absolutely required. The spindling-uniqueness problem discussed earlier is obviated by the fact that $\triangle$x and $\triangle$y fix the ends of the hypotenuse-tangent. Nevertheless, even there, the Leibnizian approach must fail, since the secants have to shrink to a point; theoretically, the minimum number of points for a secant cannot possibly be less than two. A single infinitesimal can "matter."

To prevent confusion, the hypotenuse-tangent of the differential triangle in the derivative calculus should be labeled a *neo-tangent* in contradistinction to the purely *geometrical* type. What made that confusion possible was ignorance about the infinitesimal.

Turning to Maclaurin's third drawing: "But if any right line ER (fig. 49), different from Et the continuation of ET, touch the arch EC, then the point E is a *double point* of the curve, and is the intersection of two arches which have different tangents at that point, or are on opposite sides of the same tangent, and in some cases on the same side of it."[191]

It is not a secant changing into a tangent either. E is the point of intersection of two different curves, HE and CE. The point E could be considered a zero point at which the points above it along EH could be classified as a positive veritable curved line and those along EC could be classified as a negative veritable curve, or *vice versa*.

What does it mean? Since there are lines within it, HEC is not a minimum angle. ET is a tangent for curve HE; ER is the tangent for curve EC. The two curves are not symmetrical. These tangents do not have to be geometrical; they might instead be neo-tangents, i.e, they could either just touch E, or go through it.

Today, the cusp is not considered to have a derivative, even though it is continuous. Colin Maclaurin treats this subject in depth. Of the figure 99 No. 1, on the left, he wrote: "When these arches [CE and EH] are on the same side of

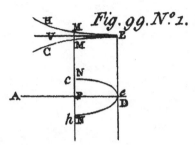

Fig. 99. N° 1.

DE, and have a tangent at E different from DE, then E is a point of reflexion, or cuspis."[192]

The tangent of the cuspis in the upper figure is EV. There is no place for another line to reach E. Beginning above V, the tangent runs by the convex side of HE, finally reaching point E. A similar course is along the convex side of CE. The tangent extends from the infinitesimal touching (or connecting with) E to V and beyond. This is the same whether the course of the tangent came along HE or CE. Since coincident lines are impossible in space, the two parts, taken separately, and the figure as a whole have the same tangent.

Maclaurin drew the lower figure ambiguously. If it is simply an arch, then the tangent is at eD. But, as he points out, if it comes to a point, then it is a cuspis, and the tangent is along Ae. The reasoning is similar to that of the upper curve. The equation for a typical form of a cuspis is $y = \sqrt[3]{x^2}$.

The cusp is not a one-dimensional spike, what moderns call a "singularity." Near its ultimate summit are the positional points. Their denominators would be below the level of the finite, and, therefore, their ratios would *also* be immeasurable infinities.

Not every imagining can exist in space. The cusp and its tangent line *can*. Contrast that with the minimum straight angle having a line impossibly touching its apex. What makes the cusp possible is that its sides curve away like the petals of a plant. The inclination of the minimum straight angle is too narrow.

The fact that such figures as the cusp or cuspis can have tangents of either kind, and *so*, legitimate derivatives, was not disputed until the full implications of the concept of the limit were applied to the interpretation of the calculus.

Some idea of the importance of the cusp can be seen from the fact that both the flower and the thorn and also the horns and tusks of animals approximate this shape.

In conclusion, since lines can exist in space, it matters where the tangent is placed. The modern teaching is that this is not very important. This brings us to a new question. The cusp has a positive immeasurable derivative on one side and an negative immeasurable derivative on the other side—a result which does not require a limit. The modernists tell us that the limit is essential; that the derivative must approach a definite limit.[193] But we have found that an immeasurable infinite is an intellectually proper answer. Wherein lies the truth?

CHAPTER XVII

THE LIMIT

$\mathbf{M}$an can attain the idea of number either out of reflection upon the discrete, the continuous, or both. With the former, the emphasis is on addition and multiplication; with the latter, on subtraction and division; the number line includes both approaches.

The goal of emancipating numbers from the spacial number line has already been shown to be a mistake with respect to the generation of the numbers themselves. In this chapter, the attempt to do this with respect to the derivative will be discussed.

The modern confusion over the tangent has come about in part because of the belief that in the theory behind the infinitesimal calculus, arithmetic, algebra, and analysis should replace geometry, the latter being relegated to a mere teaching device.

Early advocates of the limit, such as Colin Maclaurin, defined the difference between it and any member of the sequence of which it was the limit as being less than any given quantity.[194] But, this difference cannot be zero; for if it were, there would be no change. Yet, there is a difficulty with this. In his definition of the geometrical tangent, Maclaurin argues that there is not a single place for even an infinitesimal to get between the tangent and the point of tangency on the curve. This suggests that there is no →0 of any change, but a definite extent.

Let us illustrate with the differentiation of $y = x^2$: First, $y + dy = (x + dx)^2 = x^2 + 2x\, dx + dx^2$; second, after subtraction, $dy = 2x dx + dx^2$. The argument is that as dx approaches zero, the second term on the right side of the equation will tend toward zero faster than the first one. This may be so, but neither can reach it, without making dy equal to zero.

Because of the difficulty of handling the limit idea, which is supposed to justify the infinitesimal calculus without using true infinitesimals, even polished teachers sometimes make mistakes. Here is one from a standard work of the last century. The author begins; "Next, we differentiate the power function $y = f(x) = x^\alpha$ at first assuming that α is a positive integer. Provided $x_1 \neq x$, we have $f(x_1) - f(x)/\ x_1 - x = x_1^\alpha - x^\alpha/\ x_1 - x = x_1^{\alpha-1} + x_1^{\alpha-2}x + \ldots + x^{\alpha-1}$, where we divide directly or use the formula for the sum of a geometric progression. This simple algebraic manipulation is the key to the passage to the limit; for the last expression on the right-hand side of the equation is a continuous function of x_1, in particular for $x_1 = x$, and so we can carry out the passage to the limit $x_1 \to x$ for this expression simply by replacing x_1 everywhere by x. Each term then takes the value $x^{\alpha-1}$, and since the number of terms is exactly α, we obtain $y' = f'(x) = d(\ x^\alpha)/dx = \alpha x^{\alpha-1}$."[195] (Italics added for emphasis).

This argument is fallacious. When the derivative is defined in terms of "$\to 0$", then it is only a statement of a potential zero, and there is always an element of $x_1 \neq x$ in it.

To escape such difficulties, the advocates of the limit idea have tried to abstract away from the geometrical. They hold that there will be no geometrical consequences as long as the symbols are clearly and logically defined.[196] This is a recrudescence of the Hobbesian notion that we know what we make. But still, one wonders: if the world as disclosed by the senses gives a result that is somehow questionable, how can the analog be better?

One of the first to attempt to separate the limit from geometry was that of Augustin Louis Cauchy, who wrote in his *Cours d' analyse*, "When the successive values attributed to a variable approach indefinitely a fixed value so as to end by differing from it by as little as one wishes, this last is called the limit of all the others."[197]

But to speak of one value approaching another implies motion. And, although the description of motion was vital to the work of the founders of this calculus, the pioneers of modern mathematics felt that the only way to avoid the paradox mentioned above was to reduce the whole thing to numbers.

One way to get away from the idea of one number reaching another was to try to redefine the limit of an infinite sequence in such as way as to make it the same as the sequence itself. As Karl Boyer put it: "Under this view, the question as to whether the variable reaches its limit is without logical meaning. Thus the infinite sequence .9, .99, .999 . . . is the number one, and the question, 'Does it ever reach one?' is an attempt to give a metaphysical argument, which shall satisfy intuition."[198] (Intuition, namely reliance on geometrical considerations, is something they wish to avoid, presumably in order to emancipate themselves from sense.)

The truth is that this sequence cannot ever reach 1, even if person does not use geometry. This can be instantly understood. Let us translate these digital nines into fractions! $9/10 + 9/100 + 9/1000 + 9/10000$ . . . . It is clear that there will always be something missing from the sum, causing it to be always different from the number 1; the sequence following nine tenths can never equal the required one tenth. The sum in the numerator is necessarily less than the denominator, unlike .999/.999. The futility is inherent in the division process.

Even if the proposed sequence of abstract nines in the numerator could be completed (which it could not), they would not add up to the vaunted limit of 1, but only to the infinitesimal directly below it on the number line.

A way in which they attempt to get around the difficulty is through the following analogy. The reader will recall the futility of equilibrating a repeating decimal with the next whole number. Once more, this denies that there can be a "next-to". They argue, for example, that 0.9999 . . . is really equal to 1. As Professor Michael Starbird put it, "we still find that idea challenging to our intuition when it appears to contradict a bias. However, 0.999999 . . . equaling 1 is really illustrating the same principle that Zeno's arrow illustrated when it arrives at its destination."[199]

What he means is that the sequence crosses over to the limit, despite the fact that the repeating nines would always come short—a fact which he labels a mere "bias." And since post-Kantian intellectuals are supposed to eschew bias and to rise above mere "intuition," they are told that they must disregard the evident discrepancy.

He even offers two mathematical arguments in defense of this notion: (A) "Here is a proof that they are equal: Let n = .99999 . . . . Then 10n = 9.99999 . . . . Subtracting, we see that 10n-n (that is, 9n) equals 9.99999 . . . minus 0.99999 . . . equals 9. Since 9n =9, n =1." (B): "Another way to convince ourselves that 0.99999 . . . is exactly equal to 1 is to think about taking the average of 0.99999 . . . and 1. If those numbers were different, the average would be between them, but the average can only be 0.99999 . . . ."[200] From this he concludes that no point is next to any other point and that the repeated decimal somehow slides into its limit, the next integer; that Δx disappears without actually becoming zero and ruining the division.

Consider first Starbird's (B) argument, since it is the easiest to answer. The conclusion that .99999 . . . . and 1 must have an average assumes the very thing which needs to be proved. If .99999 . . . . and 1 were right next to each other, there would be no average, either. It would be exactly like the attempt to take an average between the two single numbers, 7 and 8, when they are both numbers of identity and therefore not capable of being split into fractions. Going back to principles, there is nothing between two neighboring points; therefore, no mediation can exist.

Consider argument (A). This argument can be applied to many repeating decimals involving only one single reiterated digit of 9. It can also be done when the first numbers are not repeating, for instance, 24.999999 . . . . This number can be used to give the appearance that it is in reality 25. This same can be "shown" with 8.99999 . . . to 9. But, if the general argument is that a repeating decimal jumps to some limit which is numerically different from it, why does it not work for .88888 . . . ? Let n =.88888 . . . . Then 10n = 8.8888 . . . . Subtracting, we see that 10n-n (that is, 9n) equals 8.88888 . . . minus 0.88888 . . . , which comes to 8. 000. Since 9n = 8., n = 8/9 or .88888 . . . , the same as before. Why doesn't this magically skip, also? 0.88988 . . . is closer to .89 than the last, but it doesn't leap to the limit either. 0.88989 does not cross to any different number. The closer one gets to .89 doesn't seem to matter.

Actually, the supposed discovery is a trick. It is based upon a peculiarity of the number nine. Going back to the original number, .99999 . . . : if one used only the *finite* digital sequence, .99999, the result could not be made. Part of the trick consists in slipping in an extra nine when n is raised to 10n. The other part of the trick is to divide by 9, yielding the fabulous 1.

Let us start, as before, with the original infinite sequence .99999 . . . = n. If we multiply by 9n instead of 10n, this gives a product of 8.99991 in genuine digits. Subtracting n from 9n, leaves 8n or 7.99992, which upon division by 8, returns the original .99999. Slipping in an extra 9 would not help Starbird's case at all; it would invalidate it. What is the greater metaphysical significance in multiplying by 10n rather than 9n? The fact that one can get two different answers for n simply by changing the factor shows that the process is spurious.

It is the closeness of some finite number like .99999 to an integer or a well-known fraction that makes one suppose that by some nudge which is not a nudge this difference can be crossed. Such theorists are perplexed by a supposed infinite sequence such as .88888 . . . which is closer to .88889 than to .89, but can never become anything other than what it is. Inconsistently, they suppose that an unlimited sequence running right up to a certain integer or a well-known fraction will cross, but one that is close to something else cannot.

The fallacy is in the way Professor Starbird handles the ellipsis. Numbers are finite and each place stands for a position 1/10th as great as the position on its left. The supposition that there is no such thing as a point adjacent to it is fallacious. A position next to 1 cannot be designated by a specific number, since it would be outside finite capacities of calculation. In terms of the finite, the position marked "1" and the position next to it are numerically indistinguishable. It is better to refer to the latter as "the position just below or above the number 1" on the line. Between them, there cannot be a finite difference. It cannot operate on that level in conjunction with 1. Between a numbers like .88888 and .88889, there is a finite difference. Yet, as we have seen, even when we connected them to ellipses and imagined that they were inexhaustible sequences, this technique could not be made to apply. It is only when such positions are discovered through some specific operation that they have any meaning as numbers and not as mere abstract considerations of the line. What is shown is not Zeno's arrow moving, but the fallaciousness of attempting numerical calculation after the ellipses have been added.

The impossibility of accomplishing what these modernists intend becomes even clearer when we consider something other than a supposed infinite repetition of a single digit across an endless number of places, but a number in which the digits are not the same. Suppose that you had an "infinite aggregate" which differed from √2 by a single digit. Now, how could you say that such and such is short or long of it, unless you already knew in advance what √2 was? We do know in principle, however, because we can construct it. And we also know that there must be a point below its upper boundary.

The reader will recall Weierstrass's definition that f(x) is continuous at $x=x_0$ if given any positive number ε, there exists a δ such that for all x in the interval $|x-x_0| < \delta$, $|f(x)-f(x_0)| < \varepsilon$. It was found that this definition is not valid for infinitesimals, since it does not account for next-to-neighbor-ness; it supposes that there will always be a smaller number found within the interval. This definition of continuity is almost a denial of the philosophical understanding. In place of absolute succession, we are presented with something popping in between two others, almost like that scene from the James Bond movie, where Roger Moore appears to walk across a pool of water on the heads of the waiting alligators as they surface.

Related to that continuity is the modern definition of the limit, which also came from Weierstrass. In Kline's articulation: *"The function y = f(x) has the limit b as x approaches a (or one also says at x = a) if for each positive quantity ε there exists a quantity δ such that when 0 < /x-a/ < δ then /y-b/ < ε."*[201] The difference between their definition for continuity and their definition for the limit is that while the former includes the value that could be the limit, the definition of the limit does not explicitly include it, but brackets it out by values close to it. One is not allowed to say, as close as one wishes, but rather for every ε there must exist a δ. Yet, the invisible hand of selection remains.

In another work, Kline explains: "Weierstrass attacked the phrase 'a variable approaches a limit,' which unfortunately suggests time and motion. He interprets a variable simply as a letter standing for any one of a set of values which the letter may be given. Motion is eliminated. A continuous variable is one such that if x_0 is any value of the set of values of the variable and δ any positive number there are other values of the variable in the interval $(x_0 - \delta, x_0 + \delta)$."[202]

As with the modern definition of continuity, the idea of the next-to is rejected. With this rejection, y must be a finite number, and b must be either finite or zero. They fail to make the distinction between that which is explicable in terms of numbers of unity—the *true* finite, and that which is below its level—the sub-finite. This means that the derivative must be finite, for the possibility that $\triangle x$ and $\triangle y$ be anything other than finite is by their concept of numerical division ruled out. No infinite derivatives for them, either. Their division must go on forever, never stopping.

Their arithmetical approach requires that $\triangle y/\triangle x$ converge to zero for any null sequence, whatsoever; that they converge, regardless of the way by which $\triangle x$ approaches 0; that all ways of passing to the "limit' must have the same result.[203]

But, how can $\triangle x \rightarrow 0$ in any of its ways become actually nothing without entering the realm of the infinitesimal? It cannot do so by division, for that is beyond human possibilities. The fact that their denominator is invariably finite precludes the possibility of its ever entering into the realm of the infinitesimal.

For $\triangle x$ to reach zero, it must decline arithmetically, rather than geometrically. In other words, since it cannot shrink itself to zero through division, it can only reach that result through a process of subtraction. But, in order for the subtraction of one quantity from another to yield zero, the two must be equal. This would mean that $\triangle x$ would not diminish by fractions of its beginning. There would have to be a gradual extirpation. Now, there are some physical processes which are like that, but few would argue that in every case, the reduction to zero would have to be complete.

But the modernists, of course, believe something different; they hold that it only *almost* becomes a nullity. They claim that "from the very course of the calculation that the result is independent of the particular choice of null sequence[204]" But some sequences approach zero faster than others. Since there can be no equality among the nth terms of these, short of the zero limit, their supposed arithmetical exactitude is just another approximation, glimmering through a fog. (Soon, it will be clear that the question as to whether $\triangle x$ merely tends toward zero or actually becomes it is beside the point; that, and also because $\triangle y/\triangle x$ is not just the derivative.)

As with the older idea of the limit, their inextinguishable quantity must exist outside of the sequence of that which it is to determine. In the succinct language of Edward Huntington: "Postulate 6. *Every fundamental segment of the series has a limit.*"

"Here a 'fundamental segment' is any lower segment which has no last element; the 'limit' of a fundamental segment is *the element next following all the elements of the segment . . . .*"[205] If there is no last element, then it must have a limit. And the same problems mentioned above must appear, albeit in a non-geometrical form. What do these theorists imagine a physical $\triangle x$ or $\rightarrow$ 0 to be? A concrete change endlessly getting smaller within itself? Or just the idea of something as small as one wants?

Huntington was willing to consider it to be a postulate. But sometimes, there is an attempt to get around the difficulty by turning it into an amphibole. In lieu of postulating the existence of a number S as the limit of an infinite series which converges within itself, it is *defined* by the name of the series itself.[206] By this sort of reasoning, the limit of $1 + 1/2 + 1/4 + 1/8 \ldots$ merely symbolizes this series. One wonders if this "2" is the same as the 2 which is the double the first member of said series, namely "1". And if it is, then since $2 = 2$, this number has an independent existence outside of the series, in which case, the difficulties return. If it is not, why call it "2", for that might lead to confusion. Why not call it "Abracadabra" instead?

And, what is the fate of the differential quotient when it has been taken over by the arithmetical idea? It is then fated to be a single number, not a quotient between two variables.[207] Sometimes, this is disguised by using Dy to stand for dy/ dx, "to guard the student against the misconception of regarding the derivative as a fraction."[208]

The reason why their theory must demand that it not be a fraction or a ratio is obvious. If we do cancel, then dy/dx = 2xdx/dx + dx²/dx. As dx → 0, so does the whole numerator, producing 0/0. It is only when cancellation is first done, resulting in dy/dx = 2x + dx that this limit result begins to look plausible. Actually, what is being done is to divide each term by dx.

It might be argued that an equation used to reach the derivative of the sine might be used here to strengthen their case, that sin 0°/0° = 1 might be used to get the theory out of the trap by enabling a finite result to come out of a division by zero. But, as already explained, this is not a flat nothing but an angle of zero degrees. This zero is really an infinitesimal marking a position in an angle, and the 1 is not a unity, capable of subdivision. Furthermore, dx cannot be made a 0 without doing the same thing to dy, making it also equal zero. But even if one forgot that and tried it anyway, it would not work, to wit: dy = 2x·1 + dx·1.

To repeat, their theory requires that the differential quotient be a single number—the supposed limiting value of a sequence of quotients. But, if it is a single number, it does not behave like any other. Let y be a function of u, and u a function of x, then y is also a function of x. Which means that dy/dx = dy/du · du/dx. The du cancels out. But if the differential quotient were a single number, then as Friedrich Waismann, who is of that school wrote, "dy and dx have no meaning at all taken separately, but only when used in the combination dy/dx. This symbol has meaning only as a whole; in other words, the laws of differential calculus deal with an indecomposable symbol which is merely written in the form of a quotient."[209]

But if this were really the case, then the internal substitution with du would be impossible. Contrary to Waismann, it is manifest that a single number, the denominator and numerator of which acted as parts of separate fractions in generating that number, is still a fraction. (The trivial exception is an improper fraction equivalent to a whole number.)

In terms of the way that it is formed, it makes no sense, other than as a consequence of the theory that it serves. Let us illustrate once more with the differentiation of y = x²: First, y + △y = (x + △x)² = x² + 2x △x + △x²; second, after subtraction, △y = 2x △x + △x².

According to their theory, when this equation is divided by △x, it ceases to be a quotient and becomes transformed into something very different. And this is the same theory which is supposed to have removed time from the differential calculus on the grounds that it is extra-logical. Which brings up these connected questions: When does the differential quotient cease to be a quotient? (a) Is it at the very moment that division is decided upon? (b) at the instant that it is completed? (c) the succeeding instant? or (d) some other time?

More difficulties arise when higher differentials are used. But this, we shall not discuss here.[210]

The plain truth is that $\triangle y / \triangle x$ is a trigonometric tangent, and the tangent is a *ratio*, a single number only after an act of division. The purpose behind the intellectual prestidigitation undertaken in the 19th century was not so much to generalize as to abstract away from the geometrical. As was shown in a previous chapter, the objective behind this kind of abstraction is not valid *per se*.

The infinitesimal calculus was invented or discovered in order to describe changes in motion which could not be perceived. Once this was treated as a mere historical accident and therefore as a conceptual irrelevance, the gate to the road of conceptual disaster was thrown open.

The need to risk this road resulted from an attempt to avoid the problem of the limit. Yet, the limit was introduced in order to escape the non-finite. How different this seemed back in the 18th century when Colin Maclaurin wrote that geometricians "are under no necessity of supposing, that *a finite quantity or extension consists of parts infinity in number*, or that there are any more parts in a given magnitude than they can conceive and express: it is sufficient that it may be conceived to be divided into a number of parts equal to any given or proposed number, and that is all that is supposed . . . . It is true, that the number of parts into which a given magnitude may be conceived to be divided is not to be fixed or limited, because no number is so great but a greater than it may be conceived and assigned: but there is not, therefore, any necessity for supposing that number infinite . . . ."[211]

His words did not seem unreasonable back then. They were afraid of the infinite, because they realized that there was some sort of misconception involved. If Maclaurin's term "infinite" be replaced by "immeasurable," and it be not forgotten that even a position must take up space, the need to move into the fog of the limit can be avoided.

CHAPTER XVIII

$\triangle$X AND FINITUDE

Consider the calculus. The basic paradigm of the derivative is the quadratic formula, $y = x^2$. Multiplying out $y + \triangle y$ means $(x + \triangle x)^2 = x^2 + 2x\triangle x + (\triangle x)^2$. This then is reduced to $\triangle y/\triangle x = 2x\triangle x/\triangle x + (\triangle x)^2/\triangle x$. If one were to cancel the $\triangle x$'s in the denominator, the equation would be $\triangle y = 2x\triangle x + \triangle x^2$. But the authorities tell us not to do this; they want us to consider the quotient $\triangle y/\triangle x$ to be a single number and not simply as a fraction. *But they cannot really mean this!* The part of the equation that they really want is the finite 2x. But that still leaves the second term. The old-fashioned limit theorists argued that as $\triangle x \rightarrow 0$, $(\triangle x)^2/\triangle x$ will approach zero much faster and $\triangle y/\triangle x$ will approach the *finite* 2x. There is some truth in this, but $\triangle x$ can never reach zero. Neither will $(\triangle x)^2/\triangle x$; if it did, then $\triangle y = 2x\triangle x$ would also be zero. This shows a fundamental difficulty with the way that calculus has traditionally been understood.

In the last chapter, the solution proposed by Weierstrass was found to be amiss. Quite aside from its difficulties, this arid solution abstracts so much away that it is doubtful that anyone could get the calculus from this supposed foundation if he or she did not already have the calculus in mind.

A direct solution is possible. What is required is the 2x which comes from dividing $2x\triangle x/\triangle x$. This finite number is the derivative of the curve, $y = x^{2:}$. There is no need for a long song and dance about delta-eks approaching zero. What then about the unselected part, $(\triangle x)^2/\triangle x$? Don't we need to get rid of it, somehow? Not at all; it is part of the trigonometric tangent. Simply reduce it to terms, i.e., $\triangle x$. It will be shown that this does not have to be ignored.

It has been common for mathematicians to argue that the algebraic formulation is more comprehensive than the merely geometrical. And, indeed, it is. But it brings about some products extraneous to the derivative. Two possibilities exist: (1) That not all the products of the algebraic process are necessary. (2)

That all these products are necessary, and describe the physical world quite as much as does the standard form of the derivative, itself.

Suppose for the sake of the argument that the first possibility were true. In that case, the algebra resulted in extra terms which were superfluous. And since the extraction of the derivative from the quadratic resulted in useless terms, the algebra brought forth something that was only in part consistent with reality. This would mean, in effect, that those who thought they could replace geometry with algebra were completely wrong, that geometry should triumph.

In answer to this, let us raise the following: suppose p stood for apples and q stood for oranges. What then would be q times p? It could be argued that an abstraction which was valid for one operation of arithmetic—viz. the adding of the apples, separately, or the adding of the oranges—would not necessarily be valid for both conjoined. But once there is cognizance of the fact that both are fruits, then the product would carry an existential meaning. Let the idea of that which they hold in common be supplied, i.e., that they are both fruits—and the act of multiplication applies. It is the understanding of this wider concept which makes it possible.

We know from Chapter XII that the algebraic ultimately comes from the geometrical and, therefore, cannot be used to contradict it. We also know from that chapter that the supposedly unwanted parts of the algebraic equation also appear in the geometrical construction; it was shown that if hh were to be annihilated, so would be hx and xh. Therefore, the first possibility is not correct. This leaves the second possibility, i.e., that the ordinarily rejected terms of the act of obtaining the derivative from the quadratic and higher powers represent objective reality.

If then the $\triangle$x's in the quadratic and the larger terms in the other powers are not conceptual junk, what are they? While dx·x and x·dx plainly become 2x, we still have to figure what to do with $(dx)^2/dx$. This is especially true with the greater powers. For the expansion of x^3, there are two addition products besides the finite derivative $3x^2$. These are 3xdx and dx^2. These cannot be thrown away, either. They do not become a nullity while the derivative alone is accepted. It is true that they are not part of the derivative. That is the province of the term $(3x^2)$.

On the other hand, *these additional terms already obtained in the process of derivation* would not be used in obtaining higher derivatives from the 2nd downward. To refer to the last example, the 3xdx and dx^2 would not figure in the determination of the second derivative of x^3. (Being a quadratic, however, that derivative would have its own additional product.) This, at least, simplifies matters.

One possibility is that they are single infinitesimals, i.e., they differ from the derivative only in terms of location. Were that the case, then, it might be argued that they were too trivial to be of any importance in physical science. "After all," the theorist might conclude, "since an infinitesimal has no finite dimensions, extra terms can be safely ignored. It is the derivative that we want."

This is the approach of Robinson and the school of non-standard analysis that he founded. According to Robinson, "In order that the standard number c be the derivative of f(x) at x_0 it is necessary and sufficient that f(x) - f (x_0)/x-x_0 $\cong$ c for all x $\neq x_0$ ....,"[212] This means that we may safely ignore the extra terms and simply concentrate attention on that to which they approximate, the derivative.

But if this were so, what of the tangent? If infinitesimals can be ignored, then what is kept is something less than dy/dx itself. Were that the case, this calculus would be always an approximate, perpetually stunted even at the theoretical level.

This brings up another question, what if the extra products were themselves sometimes finite?

That they are often finite can be seen from the following:

(1) Suppose, the non-derivative product is something like $\triangle x^{1/3}$: With a trivial exception, this could not be a single infinitesimal, since, by definition, the infinitesimal is the smallest there is; and this would be a small part of that—an evident contradiction. So, these extra algebraic products are almost certainly finite. Although they do not tend to zero as with the usual formulation, they are quite small. The only alternative in this case is that $\triangle x$ is a string of infinitesimals below the level of the finite. But then its root would not only be rational but also integral; fractional and irrational parts of an infinitesimal are impossible.

This is the trivial exception: Since an infinitesimal has no parts, its cubic root could only be equal to itself, as the cubic root of the one of identity is really just that $\underline{1}$, or as the $\sqrt[3]{0}$ is still 0. In all such cases, the radical would not produce a change. It would, of course, be formally consistent with the rules of algebra.

(2) Another instance would be when there was a fractional coefficient, such as 1/2 $\triangle x$. Here, it is clear that something exists which is smaller than delta-x. The latter, therefore, could not be a single infinitesimal, since that, by definition is the smallest. Such a term would either have to be finite, or would represent a string of infinitesimals of an even number. A string of odd numbers would be impossible. It is more likely that it would be finite.

(3) Consider a term like $\triangle x^2$ in which delta x has been raised to some integral power. Infinitesimals are not capable of being divided; they have no extent to them, only a location. Like the cubic root situation discussed in (1), squaring a single infinitesimal would a mere formality. Just as $1^2 = 1$, the square of a single infinitesimal can only be equal to itself. A string of infinitesimals might be squared: if $\triangle x$ equaled a change of two

infinitesimals, that squared would be four infinitesimals; there could never be a situation where delta-x squared was two infinitesimals, since that would be breaking the unbreakable.

(4) Consider again the full derivative of x^2: $dy/dx = 2x + dx$. It might seem that whether this extra term is finite or infinitesimal could not be determined in advance, but would have to await some empirical evidence in each case. Now, reflect upon the actual value of $\triangle y$, which is $2x \triangle x + \triangle x^2$. The $\triangle x$ is removed by division when $\triangle y / \triangle x$ is formed. It follows that if $\triangle y$ is finite, the $\triangle x$ which remains after this division should be finite. This is because the most that the second term on the right could be reduced to would be its square root, which should be a finite $\triangle x$. And if, *improbably*, that were infinitesimal, the derivative would be an immeasurable infinity.

Furthermore, if $\triangle y$, but not $\triangle x$, were infinitesimal—whether multi- or singular—$\triangle y / \triangle x$ could exist. A sub-finite numerator can be fractionated by a finite denominator. But it would require multitudinous changes in the denomination for every change in the numerator. Compared to it, the speed of a turtle would be above that of a modern space-rocket.

And finally, if both $\triangle y$ and $\triangle x$ were infinitesimal, their finite derivative, 2x, would be constituted strictly of numbers of identity, never of unity. No irrational numbers would work. Finite velocity and acceleration would become nearly incalculable.

But even in physical processes where the individual delta-x's are only multi-infinitesimal, the accumulation throughout the millennia would have to be finite. Indeed, one does not even have to consider all that to come to this conclusion. The mere fact that people can detect natural changes means that those which were infinitesimal in their origin had already reached the finite. Otherwise, people could not be aware of them.

(5) These important cases having been considered, let us next consider the derivatives of trigonometric functions.

    (a) Consider the full derivative of the sine: The derivative from the standpoint of geometry is the cosine. But that is not what calculus comes up with. For $y_0 = \sin x_0$, the full derivative from Calculus is $\triangle y / \triangle x = [\sin x_0 (\cos \triangle x - 1)] / \triangle x + (\cos x_0 \sin \triangle x) / \triangle x$. [213]

        According to the limit approach, as $\triangle x \to 0$, $(\cos \triangle x - 1)] / \triangle x \to 0$, leaving only $(\cos x_0 \sin \triangle x) / \triangle x$.; but since, as $\sin \triangle x / \triangle x \to 1$ as $\triangle x \to 0$, all that is left is $\cos x_0$. [214]

        $\triangle x$ can never be zero without abolishing the change. They must always be something. There is no doubt that the extra terms still exist after the change. The remaining question is whether they are all necessarily infinitesimal, or can be finite as well. The answer comes from the fact these functions often are finite. As such, some of these delta-x's can be expected to be finite. This means that the first term is indeed very small, and the second is slightly smaller than it would be for $\cos x_0$ alone.

Since the derivative of the cosine is similar to that of the sine, the same applies to it.

(b) Consider the derivative of the tangent, $y = \sin x / \cos x$. The resultant expanded derivative would vary.

(6) Consider the derivative of log $\triangle$x. The full derivative is $1/\triangle$x log e. This last part, log e, is gotten rid of by using exponential logarithms, in accordance of which log e =1. This makes the original derivative $1/\triangle$x. But its invisible presence is still there.

(7) In its turn, e is an irrational number which can exist only on the finite level. The derivative of e^x is itself. It has no extra products, but it cannot exist as a sequence of infinitesimals. Logarithms to the base e must be finite, because of this.

(8) A maxim: Any time a $\triangle$- change is or has an irrational number(s) in its neighborhood, it *must* be finite

It is established that $\triangle$x is often finite, that there is no need for an approach to zero in those cases.

The calculus originated from certain problems in astronomy, physics, and engineering. What is the application of such an insight there? It might be argued that many phenomena which have been ascribed to friction or insecure apparatus, or errors in measurement, are actually due to the effects of these insignificant extra terms.

This thesis, however, should be rejected. An insecure apparatus generally reduces the derivative's effectiveness. The tangent is final. And these tiny extra terms cannot subtract from the NET physical change provided by the derivative. They belong to the tangent.

In his study of the calculus, first published in 1742, Colin Maclaurin was able to show the geometrical stability of the derivative. They follow logically from the problems proposed; this, he showed in case after case for hundreds of pages. Friction and measuring errors deviate too much. Late in the last century, this has been shown in an original way by the Armenian mathematician, Mamikon A. Mnatsakanian. Commenting on this approach, Tom Apostol, Professor Emeritus at Caltech wrote: "As a teacher of calculus for more than 50 years and as an author of a couple of textbooks on the subject, I was stunned to learn that many standard problems in calculus can be easily solved by an innovative visual approach that makes no use of formulas. . . . Mamikon's method has some of the same ingredients [as the founders of the calculus], because it relates moving tangent segments with the areas of the regions swept out by those tangent segments.'[215]

The expansion of an expression as $(x + \triangle x)^2$ logically results in terms other than the one which is converted into the derivative. If the standard form of the derivative expresses the full change in motion, then these extra terms must refer to part of a physical action which exists, but has been overlooked, i.e., that the effective force is different from the applied force.

More pointedly: the calculus concerns itself, primarily, with the measurement of changes in motion. This measurement is traceable to the number line, which is ultimately about space.

In this chapter, we have learned that there are situations which are not fully accounted for by the derivative. The algebra is not only fully consistent with the geometry, but reaffirmed by it in the cases of the quadratic and the cubic. We have every reason to conclude that they can exist in space. What are they? Why should they continue to be ignored? The answer to these questions are important for the physical science of the future.

Let us consider a possible answer to the question as to why the rejected parts could exist, but not be detected by us.

The identification of inertial forces, unexpressed in the standard form of the derivative, follows from the Galilean principle of the independence of motions. The law of inertia requires that a body under no forces moves not only with unchanged speed but also with unaltered direction. Force is required to alter the direction as well as the speed of a body. In the case of an oblique force, velocity will be generated in the direction of this force. The connection between this principle and the ignored products of the act of obtaining a derivative can be seen in this statement by the great 18th century French mathematician, Joseph Louis Lagrange: "We understand by force the cause, whatever it may be, which impresses or tends to impress a motion on a body to which we suppose it applied . . . . It should be measured by the quantity of motion impressed or ready to be impressed. In the condition of equilibrium, the force produces no actual effect, it produces only a simple tendency to motion."[216]

M. Lagrange has been criticized for writing this. As one critic put it: "How can a force be present and produce no 'actual effect'? And in that case why should it be present? . . . . But if there is no effect, how can we study it? How can we even know that such a thing, 'whatever it may be,' is there?"[217]

To obtain a reasonable hypothesis as to what it is can be, let us re-consider the classic cases of momentum transfer through the collision of two perfectly elastic balls of equal mass, (A) and (B). Suppose that (A) is moving in the direction of (B) at +1 unit of velocity and that (B) is not moving with respect to (A). After collision, they would exchange velocities. At the time of impact and compression, t_1, (B)'s velocity would ultimately change from zero to positive one. At t_2, (B) would *react* with an equal and opposite force on (A), rejecting the latter. Action and reaction are not simultaneous, but successive. There is no reaction to the reaction coming from (A). At t_3, the two balls would be disconnected; they would not be in a state of stress. A velocity of -1 would be superimposed upon (A)'s velocity of +1, producing a relative motion with respect to (B) of 0. (B)'s relative velocity would go from 0 to +1. Thus, there is a velocity change in both balls, preserving the original relative momentum of positive one.

Note also that the reactive force produced by (B) in the direction opposite to its own impelled motion cannot be classified as a change in direction within this ball, yielding a second change in velocity for (B); a change in direction means ipso facto, a change in velocity. The reactive force is purely repulsive—no change is produced in the velocity of (B) originating from within. Suppose the contrary, i.e., that the reaction force was not purely repulsive. Were that the case, then (B), the body exerting the reaction force would either move in the direction of its counter thrust, or react against itself. If it moved in the direction of the other body, its net velocity would be less than +1, perhaps +3/4. If it reacted against itself, then its final velocity would also be diminished. A similar situation would exist for the other body.

Suppose that once (A) had separated from (B), it did exert a reaction force in the direction of (B); if it moved in the direction of (B), its net velocity would be greater than 0, perhaps +1/2. And the end velocity of (B) would be greater than +1. Momentum could not be conserved. This result being contrary to even the semi-elastic balls of experience, the conclusion that reactive force is purely repulsive stands.

Consider next the case where (A) with a velocity of +1 unit and (B) with a velocity of -1 unit, move toward each other. As a result of the collision, there are two simultaneous changes in direction in each. First, (A) and (B) cancel each other's direction; second, each exerts a force of ± 1 against the other, producing in each a reversal of direction but not of speed. (A)'s final velocity becomes -1 and (B)'s becomes +1. This means that there are two changes of velocity in each ball; the first from + or - to 0 and the reverse from 0 to - or +. This is because zero is the absence of direction as well as quantity.

Is the reaction force the only inertial force? Reaction forces must be purely repulsive. In the case of these idealized balls, there is also an elastic force which completely restores them to their original state after compression as been completed. Elastic forces are restorative only. A restorative force in a perfectly elastic ball has no internal resistance that it needs to overcome. (In a perfectly inelastic collision, there would be no restorative forces at all.)

Consider now the semi elastic objects of experience. In dy/dx, for instance, the extra $+\triangle x$ needs to be cancelled by an opposing displacement of $-\triangle x$. Let us hypothesize that within such objects there is a small capacity of perfect elasticity, a restorative force that is inherent in matter as such. After a displacement of delta x in a certain direction, there would be a restoring displacement of delta x in the opposite direction. It would, in turn, exert a reaction force against the body impressing against it, which is cancelled by the same elastic force within the other body. Neither body is strained; in order for strain to exist, (A) would have to assert an even greater force against (B). This hidden change, together with the derivative, would exhaust all the possibilities inherent in $(x + \triangle x)^2$, for instance.

If $\triangle x$ is an expression for change in time, then $\triangle x$ is the time for the displacement to take place. Including both this and full elastic recovery, the duration is $2\triangle x$.

Sir William Hamilton showed that the force corresponding to any variable is the sum of two parts, $F_r = dp_r/dt + \partial T_p/\partial q_r$ where the momentum p_r and the force F_r belong to the variable q_r, and T_p is the kinetic energy.[218] In the instance of the unexpressed elastic force, the first term, the rate of increase of the momentum of the variable with regard to time would be zero and the second term, the rate of increase of the kinetic energy per unit of increment of the variable, would also be zero, leaving the force a net zero.

With this, the $\triangle x$ of the quadratic—as well as those of higher powers—is given objective existence. It is part of the $+m\triangle x/\triangle t$ that is cancelled by the -$m\triangle x/\triangle t$ of the perfectly elastic restoring force inherent in matter. *$\triangle x$ is not some vague chimera which somehow tends to zero without destroying the standard form of the derivative, but is the real minimum displacement or associated property inherent in the perfect elasticity found in matter as such, differing in application according to the mass and/or possibly the configuration[219] of the material body being considered.* Investigation of the implications of this statement is a task for the physics of the new century.

This then is the hypothesis: that the extra terms obtained in the derivative process in calculus represent the change which is cancelled by the perfect elasticity inherent in matter.

The reader may consider this to be a standing invitation to find a better solution to the problem of the rejected terms of the derivative. To stubbornly ignore them is to doubt that algebra, geometry, and the infinitesimal calculus are reliable. Indeed, since the extra terms belong essentially to the tangent, if they had no counterpart in nature, this would be a very remarkable fact.

(For a possible application in the temperatures near Absolute Zero, please see Appendix.)

CHAPTER XIX

THE INTEGRAL

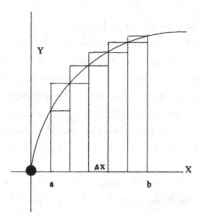

The first thing one learns when one begins to study the integral is that it is not simply the reverse of the derivative. If it were, then the original curve would be the same as the integral. Using once more the example of the quadratic, this would mean that starting from the tangent 2x, one would arrive at x^2. That would be what Newton and Maclauren called the "fluent," or in modern language, the "anti-derivative," also known as the "primitive function." The integral is different. It is the area under the curve.

The second surprise is that the typical way in which this is explained makes little use of tangents. Instead, we are given an analysis in term of rectangles. The area under the curve is to be determined by two sets of rectangles upon the same base, one larger than the area under the curve and the other smaller than the curve. In the illustration, two ordinates, one upon a and the other upon b, establish the area under the curve. The top side of the upper rectangle is determined by a line parallel to the base, as is the lower rectangle. *The tangent is missing.* The plan is to constantly reduce the basal width, making the upper and the lower rectangles come closer and closer together until they approach a certain number, called the Limit. If S_{un} is the sum of all the upper rectangles and S_{ln} is the sum of all the lower rectangles, then both approach the same limit S_n. Or, to put it differently, $S_{ln} \leq S_n \leq S_{un}$

The bases of the rectangle are supposed to shrink to a minimum. This, of course, involves the old problem of division. Under this theory, as the base shrinks to zero, this means that, ultimately, in place of a width, there will be two points. If it were finishable, the result would be two ordinates next to each other with no finite width to separate them. That would not in any case be a rectangle. Furthermore, there would be an infinitesimal difference in the length of the two ordinate lines, producing an asymmetry, albeit below the level of the finite.

Because the integral is not the reciprocal of the derivative, one should not be surprised that there is not a straight transference of dx, dy, dz, etc. across.

This scheme was originated by Leibniz[220], who, as it turns out, defined the tangent as a line through two points, infinitely near each other.[221] This is the rectangular approach's true relationship to the tangent. But two points that close do not make a line. (Leibniz' idea of the reaching the tangent through successively closer secants, we have already refuted.)

How is it that the rectangles shrink to a couple of lines? It is generally considered to be through division; as the number of rectangles increase, the length of the base gets reciprocally smaller. But since such a process is inherently unfinishable on their theory of numbers, the rectangle must always have a certain size. No specific limit that is between the upper and lower rectangle can be calculated by this method. It cannot even be calculated in theory, because the range of possible intermediate values would differ with every proportionate reduction of the two rectangles. There would always be more than one choice. Two different series of proportionately declining rectangles are set up for each curve, which, by the nature of the situation, cannot be completed. It is a potential infinity. At any given stage of its advance, we are presented with two answers very close to each other. Since the process cannot theoretically end, the rectangles do not properly calculate it. There is no "passage" to the limit.

At any given time in the devolution of the pair of rectangles toward the goal of becoming a pair of lines, the discrepancy between the areas of the upper and lower rectangles allows for a whole range of values for the limit.

And what would it be if the limit were reached? 1st, $\triangle$x could not be finite. At the same time, $\triangle$x could not be zero, for if it were, integration would be impossible. It would have to be two infinitesimals next to each other. 2nd, there would be two vertical lines next to each other of infinitesimally different lengths— at least one of which must be an irrational magnitude. Since this difference between the two ordinates stemming from each base is significant across the innumerable ordinates composing the area, chopping off this difference in order to obtain rectangles would only produce an approximate calculation—even on an ideal basis.

Yet, the calculation of the area under the curve is correct; the integral does exist. The rectangular theory as to how it is obtained is therefore wrong. This is why the dominant schools of the last century and a half went from a geometrical concept of the limit to an arithmetical idea, hoping that a full exactitude can be found by that method. At first glance, one might think it strange that they would want to use the arithmetical way of proceeding: under their theory, as we have seen, a line has as many points as an area. Applying it to the integral, one could conclude that the linear distance between a and b on the x axis has as many points as the area above, which might lead one to abandon the integral altogether in favor of the anti-derivative. But this is not what they do. Instead, as we shall see below, their new way is really the old way stated in such a manner as to follow the course of the former, while denying it through the device of a strange terminology. And by doing so, they only succeed in continuing its errors, while compounding the confusion.

Let us at this time consider a geometrical analysis of the integral based upon the true nature of the tangent, i.e., as a line that connects with the curve at one

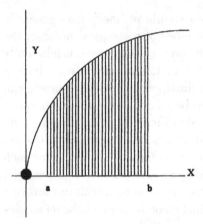

point, not two. Once the two bonding ordinates under the curve a and b are selected, the area automatically exists. No rhythmic approximation can account for it.

Consider any point on the curve which the tangent meets. This point is the top of an ordinate, in this instance, y. This ordinate is x distance from a selected origin. The area under the curve consists of an immeasurable quantity of these ordinates next to each other. It has already been shown that when the lines cover every point, they must comprise the entire area. Recall the area of a square which was covered by successively laying lines of equal length next to each other. With the curve, the lengths of the ordinates vary. The relationship between the y-length of any given ordinate and its x-distance from the origin establishes the unique situation measured by its respective tangent. This is the case with each of the infinity of ordinates. The formula is perfect, and there is no passage to a limit external to what is being calculated.

The geometrical tangent is a line outside the curve which measures the slope at that point; it is not part of the curve. Its use is compatible with the kind of curve pictured above where the convex side faces away from the base. But in curves like the exponential and the parabolic section where the opposite occurs, the neo-tangent is better, i.e., the tangent line should begin *in* rather than *on* the curve. It is better, because otherwise, a zero point for a curve that was in one sign would have started its tangent in the other sign. As far as this writer can see, those interested in uniformity could use the neo-tangent for both.

Under the rectangular approach, the limit is a point on a curve f(x) which is approximated as $\triangle x \rightarrow 0$. But if $\triangle x$ were really 0, there would be no change. Therefore, delta x cannot be zero. Moreover, it is incorrect for them to speak of one point on f(x); in strict consistency, their theory calls for two points. The escape through the limit is not successful.

The tangent method necessitates that the level of the infinitesimal be reached; since there is no finite width to an ordinate; the finitude is in the length of the ordinate. This method is direct. In contrast, the other method must treat its limit—the area—as something outside the method of calculation, which is the diminishing pair of rectangles resting upon the same base. It is indirect.

A critic might argue that the tangent method has the following problem: $\triangle$x, is usually finite, but the difference between point P and the next point on either side is supposed to be infinitesimal. How can a finite change be based upon an infinitesimal difference?

The answer is that in one respect, it cannot, and in another respect, it can. The change being measured by men is finite, whether it is velocity, acceleration, or whatever—depending upon the nature of the reality being considered. The usual measurement taken by men when they operate the calculus is not point to next-point, but from one point to another point a finite distance away. In order for a change to be subject to the human scale of measurement, each of the points must be finite with respect to some other point on the curve.

Yet, each and every point on the curve is an interval apart from another point a finite distance away. Hence, a finite distance exists somewhere for all. This calculus requires awareness of infinitesimals, for the ordinates need to follow one right after another, neighbor by neighbor; if they did not, there would be a breach in continuity. Yet, the only changes men can measure are finite. With the rectangular method, not only must the changes being measured be finite, but limit is inexact.

Integration does not fail. However tortous the curve, the area under it with respect to the base exists, and the ordinates composing it completely fill in the available spaces. However finite the $\triangle$x's which constituted the change may have been, to the extent that the path traced by their accumulation is solid, an exact integration can be made. (The same is true for volumes.)

The tangent method is perfect. Since it can consider a difference of one x-position from its neighbor along a curve, its theoretical range cannot be improved upon. The theoretical minimum for the rectangular method is two points; it is off by an extra position.

Usually, the indefinite integral is expressed as $\int f(x)\, dx$. But integration does not proceed by a finite dx getting smaller, but by ordinates of various lengths, laid next to each other, filling up the area under curve. The corrected indefinite form should be $\int f(x) \bullet_x$. A double integral would have two subscripts after the infinitesimal sign, such as $\int\int f(x,y) \bullet_{xy}$; a triple integral would have three, for instance. $\int\int\int f(x,y,z) \bullet_{xyz}$. The tangent representation is in exact conformity with the truth.

Abraham Robinson's non-standard analysis agrees with the tangent thesis in that its partition of the area under the curve should be analyzed into an infinity of partitions, but it tries to use an infinite natural number, ω and also resorts to approximates on the level of theory.[222] But ω cannot exist, since every number is finite and every theoretical approximation is inexact by definition.

To this, the advocate of the limit approach might answer that if areas and volumes are handled by the tangent method, arc lengths are better handled by way of the limit. Its indefinite integral s = lim $\sqrt{[1 + (\triangle y_k/\triangle x_k)^2 \triangle x_k]}$ where n→∞ and k =1,2,3,4 . . . . Since $(\triangle y_k/\triangle x_k)^2$ is the chord squared, the limit of the sum of these chords (which are in fact secants) is the length of the arc.

The answer is that it is actually an approximation formula, but can be made as exact as the person using it wishes. Any curve of any kind, like the straight line, is capable of innumerable divisions. The true arc would be greater than the length of these secants. But since straight lines and curves are not the same, this formula does not give an exact amount. Nonetheless, this interpretation is used to explain the correct answer by means of the familiar "limit" rubric. Three things should be said about this. First, it does not use the rectangular method either, but progressively smaller *secants*. Second, there is a well-known trigonometric tangent equation for the integration of the arc which provides the right answer. This brings in the differential triangle, and, therefore, the neo-tangent. (There is also a relationship between the trigonometric tangent and the secant, but not with the geometrical tangent, *à la* Leibniz.) Third, the tangent method can handle approximations, also. There is no need to resort to an inferior approach.

Another critic might ask: what happens on the range near zero below the level of the finite? If there are no lines from base to curve in this portion of the area, how can an accurate determination be made? Wouldn't this be an approximate also, a case of a theory not quite meeting the requirements of exactitude?

This critic will be successfully answered in Chapter XXII, when the subject of the stability of the linear unit is addressed.

Now, let us return to a comparison of the two accounts of the integral. The people who use the rectangular approach agree that theirs is inexact. They attempt to replace it with an arithmetical approach, but the latter is really dependent on the former. Kline, for instance, offers the following theorem as proof that his sequences which are originally drawn from the rectangles have limits: "*If f(x) is continuous in the interval $a \le x \le b$, then the definite integral . . . exists.*"[223] He argues that it means that both the upper and lower sequences lead to the same limit. The significant part of his argument is as follows: "However, the geometrical area has no precise definition, whereas the definite integral does have a very satisfactory definition. Hence we *define* the geometrical area to be what the definite integral yields."[224]

This is the difficulty with the whole procedure. How does one know that this "area" which he claims he cannot define precisely will fit into this concept?

How does he know that the two will be congruent? He hopes he knows because he has carefully defined his arithmetic argument so that it follows the line of his rectangular argument. The arithmetical argument is only an analog. It seems to stand on its on feet only because he has already found $S_{ln} \leq S_n \leq S_{un}$ to be eminently reasonable through its rectangular guide. If it were not, he would not dream of defining it as "what the definite integral yields."

This is the problem, not just with Kline, but with that whole school of thought. Henri Lebesgue wrote concerning the unbounded interval, that "it would require an infinity of numbers y_1 to divide it into intervals of length at most equal to ε."[225] If this ε is on the x-axis, and if it has any finite size at all, there cannot be an infinite number of ordinates. If it is on the y-axis, it does not work there either.

And the reason for switching to the arithmetical is that the rectangular argument cannot get beyond the finite. Yet, the modern Weierstrassian idea of continuity is tied up with the Weierstrassian ε/N idea of the limit where, in Kline's words, *'the number S is the limit of the sequence $s_1, s_2, s_3 \ldots, s_n \ldots$ if given any positive quantity ε, there exists a N such that for all n > N, $| S > s_n | <$ ε."*[226] This idea is not free from the difficulty of treating the small-as one-wishes as if it were an infinitesimal; ε is still finite. Those who wish to avoid any recourse to the infinite suppose that such considerations will lead to some specific limit when it cannot. Their supposition is based upon a hope that the resulting blur will somehow serve for a passage to the limit.

What is required is that the operation proceed from point to point on the curve, from neighbor to neighbor, an operation which that definition of continuity simply cannot provide—cannot provide because its very premise is that between any two numbers, a third can be found. In actual continuity, there must exist two points between which a third cannot be inserted. The tangent method can handle both point to point and finite changes.

Just because we can generalize a geometrical result with algebra does not make it automatically valid, as was shown in the last chapter. And while most of the derivatives involve finitude of change, all of the integrals consist of infinitesimal change. This is not paradoxical, for the inverse of derivation is not integration, but the fluent or original function.

The unreality of the limit approach can be also be found in their handling of the improper integral; this type of integral concerns situations in which there is a discontinuity somewhere. A simple example would be a function which has a y value of ½ between x values of 0 and 1 and then jumps to a y value of 2 between x values of 1 and 5. This is solved by adding the two parts. Yet, where approximations lie in any term, the difficulties remain.

At the theoretical level, higher improper integrals contain additional errors. These integrals use a modified version of the rectangles, which is usually termed "partitions." The Riemann integral use of the notion of oscillation requires a movement back and forth of the variable above and below zero as determined by the upper and lower approximates as the variable approaches its limit[227]; the Stieltjes integral involves a slightly more complicated handling of the limit[228] and succumbs to the same mistakes as have already been discussed.

The Lebesgue integral adds a new fallacy with its notion of the "measure," especially in its handling of so-called denumerable and un-denumerable infinities. Space does not permit a further discussion of it in this place, but the reader can find a brief discussion of it in the footnote.[229]

Substantial criticism has been made of Leibniz. Yet, strangely enough, he was an exponent of the infinitesimal. However, he did not regard them as existing in reality. As he wrote to Johann Bernoulli, who really did believe in infinitesimals: "For perhaps, the infinite, such as we conceive it, and the infinitely small, are imaginary, and yet apt for determining real things, just as imaginary roots are customarily supposed to be."[230] Many other statements of his show that he held them to be fictions. As he put it to another correspondent: "I do not dispute whether these inassignable quantities are true or fictive; it suffices that they serve for the abbreviation of thought . . . ."[231]

Many scholars hold that he had a hidden philosophy quite different from the one in his public writings. They think the two sides of him are reconcilable through his doctrine of "harmony," wherein contradictions are ultimately reconciled. For instance, Leibniz did not believe in causality, but that the flow of events was such that it would seem that A caused B, whereas they only followed each other. In his mind, infinitesimals did not actually exist, but the structure of the realm of mathematics was such that it seemed as if they were real.[232] In this regard, he was an ancestor of Hume and Kant.

The infinitesimal does not need friends like that.

CHAPTER XX

NON-EUCLIDEAN GEOMETRY

Having proved the existence of the infinitesimal in Chapter VII and having subsequently apprehended some of its workings in important areas of thought, let us look at a popular conceptual aberration, the notion that space can be curved. That this is inconsistent with the infinitesimal is obvious. Any curvature must, by definition, consist of more than one location. But locations are indivisible; therefore, no location can be curved. Space is neither straight, crooked, nor curved; rather, it is that into which curves fit.

This being understood, we find ourselves in conflict with one of the reigning ideas of mathematics of the last hundred and fifty years. Yet, we do not have to be very brave to oppose, since during this time the science of mathematics has lost the claim to certainty it once had.

Non-Euclidean geometry was imagined early in the 19th century. Many of its devotees, including Gauss, were interested readers of Kant. Kant had taught that space did not have to exist in external reality, but rather that it was the way that the mind necessarily organized experience. He thought that our minds must use the geometry that he knew, what is today called "Euclidean geometry." But some of his followers saw that if the organizing principles were a little plastic, some changes could be made, and a world just as valid as the one taken in by the senses could be projected; that experience was essentially relative. A case in point is Bernhard Riemann's idea of magnitude, which was that "notions of magnitude are only possible when there is an antecedent general concept which admits of different ways of determination. According as a continuous transition does or does not take place among these determinations, from one to another, they form a continuous or discrete manifold; . . . individual determinations are called points in the first case, in the last case, elements of the manifold."[233]

In his book, *The Labyrinth of Thought: A History of Set Theory and its Role in Modern Mathematics*, José Domínguez Ferreriós, states that Riemann was using the words "manifolds" and "magnitudes" as synonyms.[234] Ferreirós adds that Riemann's "notion of a manifold, so defined, does not depend at all on spacial intuitions. The spacial notion of space, . . . and line are only the simplest, intuitive examples of three, two, and one-dimensional manifolds."[235]

It is clear from the foregoing that Riemann did not mean by this that the "manifolds" come from the senses. They are Kantian transcendental concepts which organize experience in certain ways. These manifolds are imagined to be more basic than the ideas of space themselves. Unlike the three dimensions (length, width, and height), they are not conceived of being abstractions from our basic experience of space; rather, the manifolds are to be metaphysically superior. It is imagined that in terms of these shadowing conceptual clouds, our experience is organized as "spacial." According to this philosophy, man does not perceive reality; he interprets it.

Our purpose, however, is not to explore Kantianism, but non-Euclidean geometry. Let us now consider many of the problems of this 19th century "addition" to geometry.

First of all, it does not appear in perception. Long ago, Ptolemy showed that dimensions must be determinate, and determinate dimensions must exist along perpendicular straight lines, and that it is not possible to place anymore than three such lines at right angles to one another.[236] People did not first discover the point, a zero dimension, then the 1st dimension, length, then the 2nd dimension, area, and finally the 3rd dimension, volume. Actually, the reverse was the case; first, they discovered space, and then intellectuals drew abstractions from that. *Flatland* is just a clever late 19th century phantasy.

Second, its claim is that if it is consistent with itself, it must stand. The historic basis of their argument will be stated below, but let us here consider the claim of mere consistency. Its wide-spread reception was due to advanced Kantian notions that experience is essentially malleable.

In Chapter XII, the case of fiction was brought up, wherein the slight differences of circumstances with respect to the same general subject—say courting and love—could produce greatly different results, owing to the enormous complexity involved. Fiction writers often speak of their premises. Yet, however logical the story line may be and however carefully interwoven the sub-plots, they are simply not real. Scarlet O'Hara and Rhett Butler never existed. More to the point, the old Walt Disney cartoons of live wooden puppets, flying elephants, and mouse-people may be as self-consistent as one likes, but they are still impossible. Their internal coherence does not allow them to break into reality.

Third, in the study of the syllogism, it has long been known that the reasoning itself can be logically valid, but both premises and the conclusion can be false.

An example of that was given in chapter VI. It follows that this can be so, not only for a single syllogism, but for whole chains or sorties of them. Self-consistency is not enough.

Fourth, it comes in part from loose thinking about the plane, which, as was shown in Chapter XII, is merely an abstraction from the practical need to have a hard surface on which to write. Beyond that, it has no metaphysical import. A plane is not part of a drawing. It only provides alignment. If what is being depicted is a circle, a flat surface simply prevents an oval or some other figure from being drawn instead. It does not define space in some hyper-intellectual way; if it defines it at all, it is only in a rudimentary, practical way. Of course, by using a sphere for a writing surface, you can draw triangles with the three interior angles greater than two right angles. But space is not curved; rather, it is that within which curved and straight and nondescript objects are possible.

Let us turn to the actual claims of non-Euclidean geometry. It has been argued that $\sqrt{-1}$ makes possible the idea of space-time. The mistake issues from an unwarranted extension of the practical definition of an imaginary line as that which exists at right angles to a real line. This works with a 2-dimensional Cartesian representation. But when applied to a 3-dimensional system of real axes, no place remains for the positioning of a fourth axis. The discovery of the veritable number and the understanding that the imaginary axis is a real-veritable hybrid corrects this mistake.

It is said that the famous parallel lines postulate of Euclid logically admits a contrary. And that this allows for a different kind of space to exist. What is forgotten here is that Euclid's words are not like a fiction writer's, but rather are drawn from experience. In summary: the non-Euclideans postulate is a pretended equality between abstractions drawn from sense and mere posits. Stated in that way, the difference in metaphysical rank should be obvious to all but the most resolute subjectivist.

But let us consider their argument.

Euclid defined parallel lines as "straight lines which, being in the same plane and being produced indefinitely in both directions, do not meet one another in either direction."[237] Those writers are not satisfied with the objection that the definition is negative; they add that they can construct, without contradicting themselves, a system in which there are no parallel lines, etc., and that therefore this construction must mean that there is a reality which is far different from the mundane, bourgeois world of three dimensions.

In fact, there is no insuperable difficulty about parallel lines at all. An adequate definition was given by the late president of Duquesne University, J. J. Callahan, over seventy years ago, this being: *"Parallel straight lines are straight lines that are equidistant at equidistant points.* For instance, the lines x and y are

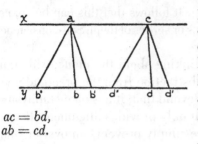

$ac = bd,$

$ab = cd.$

parallel if, taking any two points on one line, as a and c, and two points, such as b and d, on the other line, so that the distance between b and d is equal to the distance between a and c; the distance between a and b, and c and d are also equal. Or ac = bd, ab = cd."[238]

The choice of points a and c on the first line and the distance between them is arbitrary. Likewise the distance between a on the first line and b on the second line is also arbitrary. The sole requirement is that the two equations hold between corresponding points. There is no need to choose the line with the shortest distance; the angles do not have to be right. As Callahan said: "Whatever way the distances are taken, perpendicular or otherwise, providing the distances are taken between corresponding points, they are always equal if the lines are parallel."[239]

The definition is not negative; it is universal. The lines are placed under the genus of being straight. Neither is there any pretense that the lines are infinite in the sense of being endless; all that is required is a potential infinity. His improvements over Euclid are substantial; they follow the original procedure, making changes only when necessary. There is no presumption of modern superiority.

Callahan showed that if one wants a geometrical situation where triangles are not the sum of two right angles and in which straight lines cannot be drawn on a surface, one does not have to look for anything more exotic than the surface of a sphere, a part of traditional solid geometry. Callahan went much farther than that, but his work was ignored, until it was reintroduced to the attention of the intellectual public by Turner and Hazelett.[240]

Euclid's definition was not final, but it is a fact that if two lines are equidistant, they can never meet. Nevertheless, the characteristic of never-meeting is not unique to the parallel. Consider the following example: Two straight line next to each other would not be parallel, since no distance separates them. But they intersect *not*. Another example: A line on a plane may be extended indefinitely, i.e., *ad infinitum*. Let a second line lying on the same plane be extended until it intersects the first one. By both the true definition and the non-Euclidean one, it is not parallel to the first line. But consider a third line which begins above the plane of the other two and pierces their plane where those lines are not. It will never meet either line. Yet, it is not parallel to either of them. The negative condition of never meeting does not necessitate parallelism.

Let us look at the claim of the non-Euclideans a little more closely. They acknowledge that when the possibility of parallel lines is excluded, quite a number of important geometrical propositions are denied, but defend themselves by saying that not all of geometry is removed. They argue that an alternative science of space can be erected upon what is left. But this is a non-sequitur. It does not follow that if one denies only a large part of something as fundamental as space that one can erect a different structure in its place that is equally as real.

Frequently, two of the most common attempts, Riemannian and hyperbolic geometry, are presented together with Euclidean geometry as if they were equals. Consider hyperbolic geometry. The initial premise is that parallel lines can be acute with respect to a line with which it is said to be "parallel." This, of course, should be rejected outright as false.

How do its advocates protect themselves from being dismissed immediately? The advocates of hyperbolic geometry argue that because their intellectual confabulation does not destroy all of geometry, it must be "consistent" with that which remains. It destroys squares and rectangles, because they depend upon parallel lines, but leaves triangles, although they can never be similar to each other.[241] It allows for some perpendiculars. To argue that it is valid because it leaves some parts of geometry intact is fallacious. It simply means that there are parts not affected by its denial. That is worse than saying that a bad law is good, if it does *not* invalidate the entire legal system. Consider the mundane case of a standard cardboard puzzle. If one of the pieces were removed, whatever replaced it would need to have the same shape and size in the aggregate. Any shape which simply could be placed within it, but would not fit perfectly every point of contact with what was left of the old puzzle would not do. Only the original shapes would work. This would not change even if one used little bits to add up to the wanted shapes. Only that which added up to the original ones would work. Anything less than that would fail by necessity. The same is true with trying to describe a space without parallel lines.

The fact that an intellectual can put together some axioms without using parallel lines as defined above is not evidence that there exists a "parallel" universe besides the one we know. One can also put together some rules in which no straight lines can incline with respect to one another at any angle except a right one, that none can cross at a different angle, that the only lines allowed besides right and curved ones are the parallel straight sort. Under such a restriction, squares and rectangles could exist, but without diagonals. This would not afford evidence of an alternative universe, but only of an impoverished geometry. It is not likely, however, that anyone would make that claim. A "universe" destitute of diagonal lines would lack the loop the loop attractiveness craved by 20th century revolutionary intellectuals.

Figures can be drawn across corrugated surfaces like an old fashioned washboard. The resultant geometrical "objects" would consists of lines bent at the same angles. Doubtless, formulas can be written for the construction of its square-like and circle-like figures; in fact, comic versions of all the conic sections are easy to imagine. But this would in no way imply the existence of washboard-like space. It only means that a washboard can be used as a "plane," or more exactly as a writing surface, something any kid who had ever tried to write on a sheet of paper placed on a rug should know.

Moreover, even if one were to mentally exclude consciousness of straight lines altogether, there would still remain in reality a parallelism of lines fully consistent with Callahan's improvement over Euclid. This would be concentric lines; they can never by definition touch each either. In such a truncated geometry, parallel lines would not be potential infinities. Since there would be no way to find the radius or diameter of a circle, π would probably never be found, and if it were, no one would know its primary use. This dim projection can be appreciated without hypothecating an alternative universe.

Much more to the modernist's liking is the exclusion of straight lines, even a diameter, through the devising of a geometry based upon the surface of a sphere. This is what Riemann did. By the same token, a talking Queen of Hearts playing card may be consistent with the rest of *Alice In Wonderland*, but that does not mean that one is going to meet some interesting characters after chasing a white rabbit down its hole.

In conclusion: non-Euclidean geometry undercuts understanding of the infinitesimal by implicitly bracketing out the latter's indivisibility. Its advocates gain acceptance through arguing that it is consistent with itself. They are able to get by with that because of the widespread acceptance of Kantian and relativistic ideas among regnant intellectuals.

CHAPTER XXI

SETS

Right from the beginning, we should suspect that there is something wrong with the idea of a "set." For instance, consider the routine definition of a circle as "the set of points equidistant from a fixed point, its center." At first glance, this seems harmless enough. But upon closer inspection, one finds that it is not universal. In Chapter XIV, it was shown that when veritable axes are crossed in the standard Cartesian form, although the requisite points exist in the 1st and 3rd quadrants, they cannot be connected to form a circle; that quarter circles exist only in the first and third quadrants. The points in the two figure-less axes (the 2nd and 4th) cannot form a line with the other two, because of mixed signs. To be universal, the definition should read: "The circle is a continuous figure, every point of which is equidistant from a fixed point, its center." A set of points is merely a collection; and that is not what a line is. Although it is made up of points, its realm of reality cannot be reduced to them without annihilating its identity. In order to exist, a circle must be constructed; the place where it is drawn is only a necessary condition for it. To ignore this is to ignore part of its reality.

This lack of universality points to difficulties more fundamental. To recapitulate: A number is a quantitative relationship between two points separated by an interval established by a standard serving as a unit. Whole numbers and rational fractions can be stated exactly in terms of this standard; irrational numbers cannot. Note that the concept number presupposes the concept quantity. Until a quantity has been identified in terms of some such standard, it is not a number.

Set theory holds that a group of numbers of a certain kind, such as an infinity of rational numbers, already exist, when, in fact, only a few are actual. Imbedded in this is the theoretic loss of potential infinity. A hunter would never say that a bird in the bush is worth as much as one in the hand; nor would a bird watcher say that the one never espied is as well perceived as the one seen. Yet, the set theorist talks about rational numbers as if they were all collected into some kind of a theoretical bag. The incalculable nature of infinity, any infinity—even the one that lies between zero and one half—is treated as if the difficulty were only practical—merely a matter of the shortness of human life, etc.

In keeping with the modern preference, set theory conflates potential and actual infinity. With the breaking of that distinction, impossible attributes are arrogated. Although the claim to possess the contents of an unattainable future is usually not made outright, set theory would be valid only if that were the case.

This can cause confusion even on the lower levels of conceptual awareness. When the members of a "set" are regarded as separate elements, one might tend to forget that in many cases they are made of each other. Consider, for example, a person who was told that the set of even natural numbers are 2, 4, 6, 8, 10 . . . . This person might forget that 4 is composed of 2, that 6 contains both 4 and 2, that 8 has all three of what went before, and so forth. It is as if one were to count the principal elements of the hydrogen atom as being three: (1) the electron, (2) the proton, and (3) the atom itself. Those conversant with the number line would never say that "2 is a rational point," since without knowledge of the point at the origin, that statement would be false. It takes two points, not one, separating a finite interval to establish a number.

Aside from breaking the barrier that separates actual from potential infinity, certain other streams running through the 20th century intellect contributed to the course of modern set theory.

(1) We have noted that many were attracted to the idea of the limit, because with it, they could carefully avoid reference to the infinite. This was the thought of Newton, Maclaurin, and others. Although it may have seemed safer at the time, this approach was not ultimately successful; the supposedly clear path got entangled up with the difficulties of the limit. Peculiarly destructive was the strategy of attempting to slowly meld the approximate with the exact; this was to be accomplished by the limit. The inventor of this dodge may have been Leibniz. In a 1687 letter to Pierre Bayle, he writes: "In any supposed transition, ending in any terminus, it is permissible to institute a general reasoning, in which *the final terminus may also be included* . . . . Yet a state of transition may be imagined, or one of evanescence, in which indeed there has not yet arisen exact equality or rest . . . ." (*Italics* added). Leibniz, however, had some reservations: "for the present, whether such a state of instantaneous transition from inequality or equality . . . can be sustained in a rigorous or metaphysical sense, or whether infinite extensions successively greater and greater, or infinitely small ones successively less and less, are legitimate considerations, is a matter that I own to be possibly open to question."[242]

Later on, these reservations were either gone, or left unexpressed. In a 1695 essay, he wrote: *"I think that those things are equal not only whose difference is absolutely nothing, but also whose difference is incomparably small; and although this difference need not be called absolutely nothing, neither is it a quantity comparable with those whose difference it is."*[243] (*Italics* added.)

(2) Another source were intellectuals who thought that irrational numbers were not neat. As expounded by Friedrich Waismann: "To know an irrational number means to know a process for computing approximately. In this sense it is completely proper to identify the irrational numbers with the approximation process (the sequence)."[244] This is a non-sequitur; the fact that the calculation of a magnitude is perpetually inexact does not abolish the difference. To identify it with its approximate is to remove that difference from consciousness.

(3) There was also the belief that there is something wrong with basing the number system and mathematics on geometrical considerations, that basing it on arithmetic alone was purer, because there is no seemingly extraneous element to which they are attached. These people tried to replace Pascal's dictum that all that transcends geometry is beyond our understanding by substituting arithmetic for geometry.[245] Strangely enough, this idea was coupled with the intent of preserving a stilted form of the axiomatic method in which the basic terms were left undefined. It is true that with propositions deduced from only a few axioms, the mathematician can proceed from one to the next with great ease. But when the basic terms are left blank, the whole apparatus leans upon the specificity of the axioms.

(4) The theory of sets depended heavily on Bernhard Riemann's idea of manifold, which held that the idea of magnitude was independent of the notion of space. This allowed numbers to be separated entirely from the infinitesimal and the number line.

These swollen currents—the contradiction within the limit idea, the hatred of irrational numbers, the anti-geometrical attitude arising from refinements of Kant, Riemann's manifolds—surfaced in the late 19th century as the standing wave of set theory. Since the numbers approaching a limit and the limit itself are all numbers, placing them in the same "set" allows the believer to suppose that a bridge exists by which the narrow chasm can be crossed; or better, that there is no need of a bridge.

With set theory, the problem of the limit is ended by naming the set after the limit, thereby supposedly including all. Thus, $1/2 + 1/4 + 1/8 + 1/16 \ldots$ are supposed to be all present in the set—the whole infinity of them included by this ellipsis—and it is supposedly the set of "1," which is, so to speak, the case enclosing the set. Similarly, .3333 . . . is called 1/3, even though the former does not equal the latter, no matter how many repetitions of the number three are made on the right side, and the "full number"—if one can speak that way— is unreachable. With 1/3 being the box into which the theoretically interminable threes are placed, the chasm can be imaged to be sutured shut.

But there is another fallacy in it even more fantastic: the theory of sets implies that in some way the instances of an idea are conjoined with the idea itself, that what is intended is treated as if it were actual. Adherents of the theory speak of the set of rational numbers being infinite, even though men have only conceptualized a finite number. By this, they do not mean that it is a potential infinity, one that is never complete, yet that their definition will apply to any in the future. No, their concept is supposed to embrace what they have not. They are able to think this, because what has been identified and what has not been identified are both placed into a set which includes them all anyway. The contrary idea that a concept merely applies to whatever in reality may correspond to it is foreign to this attitude. For the set theorist, it is not enough that the concept's definition have an application; the concept must include its application as well.

To understand why set theory makes this seem not to be an illusion, it is instructive first to look at an old theory about the nature of concepts. This is the traditional division between intention and extension. As a standard work on logic puts it: "The **intension** of a term is what we *intend* by it, or what we mean by it when predicating it of any subject: the **extension** is all that stands subordinated to it as to a genus, the variety of kinds over which the prediction of the term may *extend*. "[246] Thus, when we use the word "word," we ordinarily intend by it an audio-visual symbol, signifying a person, a thing, a place, or an idea. Under this rubric, the intention was basically the idea itself, what is meant by it; the extension consists of that to which the idea applies. Thus, the intention of circle is its definition that it be a closed line, all parts of which are equidistant from a center point, while the extension consists of all the circles that actually exist. Similarly, the intension of the word "book" is its definition, while the extension is the totality of books altogether.

One trouble with this theory is that, unless one is very careful with it, one might think of the referents of the concept as somehow conjoined with the ideas used to identify them. People who use this theory speak of the referent being "contained" in the concept or of the concept "comprehending" its referents; the mental image is that of the concept being a great box into which the referents are included. This tends to produce a psychological idea that the mind somehow includes all of reality. Most of the logicians who embrace extension-intension theory are careful to avoid this trap, claiming that extension and intension are really just two ways of referring to the same object. But, for the careless, the miasma is not far away.

Ferreirós points out that set theory is both intensional and extensional.[247] Speaking figuratively, the concept of a set is shaped like a bar bell with intension at one end and extension at the other. This give the concept two separate centers of focus. One is never sure whether the parts of a set are all considered to be fully mental or not.

The easiest way to get a glimmer of understanding about this concept with

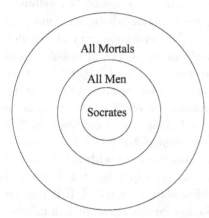

two separate points of focus is through a near ancestor, this being the Venn diagram. These diagrams show the relationship between the intention and extension in a general way. The major premise of the syllogism, "All Men are Mortal," is pictured as a circle called "All Men" placed inside another circle, called "All Mortals." The minor premise, "Socrates is a Man" is pictured as the circle "Socrates" placed inside the circle "men," which leads to the conclusion that "Socrates is Mortal," since it can be represented as a circle inside a circle inside a circle. Here, the extension is illustrated in the shape of a circle and intension by what is written on it. One can see how the dramatic character of this mode of illustration might make it possible for someone to think of the two as somehow the same. But they are not. The circles merely illustrate a syllogistic form, in this case the categorical affirmative. The qualitative subjects and predicates of each of the two propositions in this or any other syllogism must be clearly identified.

Most likely, all men have not been born; yet, the circle casts the totality as a collection, even though all cannot be present. This mode of presentation distorts the true meaning of the syllogism. It is not by virtue of belonging to this group that a man or woman will die, but because of his or her possessing the attribute of mortality. Even if there were but one human being and that person had not eternal life, the fact that he or she will someday die would be a necessary conclusion.

The Venn diagram was identified early in the last century as a crude and naive nominalism.[248] When used on quantities which are potentially infinite, Venn's circles really crack. What is not there is treated as if it were. That to which a concept pertains, the reality which it identifies, is not something set within a circle formed by the concept. It does not "embrace" them. Unlike a geometrical tangent, it does not even touch them. Rather, it *refers* to them.

Early in the 20th century, certain advocates of set theory became aware of some contradictions in it. For instance, does the set of all sets include itself? This is because of the box-like structure of the idea. How can a box include itself as well?

Bertrand Russell tried to handle this with the division of sets into those which include themselves and those which do not. While this escapes formal contradiction superficially, it is really quite unnecessary; it makes the idea of a set even more mysterious. Conceptual theories based on reality are not compatible with set theory. Presented with an idea like "everything," the realist would not be concerned about the question as to how the idea of everything must also pertain to the holder of the idea and the idea itself. The concept, "everything," is not a box containing all that is. Consider the situation where one concept is the genus of the other. The genus "drinking cup" is not a compartment in which may be found china cups, porcelain cups, plastic cups, etc. It is but a word which pertains to any material object used to contain a liquid for drinking. They can be classified, according to their material composition (as above), or according to their size, or country of origin, or in a number of other ways. There is no confusion if a porcelain cup is also classified as having come from England. The concept *does not touch* these objects; it *refers to them*. What they refer to is not a collection, not something in a box. This is also the case with mathematical concepts: the genus "quantity" includes as its principal species, finite and infinite. Quantity is not a giant box into which both are packed. If it were, it would lead to such bizarre questions as to whether quantity is more than infinity, because it "contains" infinity as well. Rather, quantity is the name of the attribute which both the finite and the infinite and their subdivisions possess.

Set theory obviates the need for induction. By placing both extension and intension within the same concept, the theorist can conflate the difference between potential and actual infinity. Such a theorist supposes that he or she does not need to abstract from reality and develop this abstraction into a basis for future deductions and inductions; that, instead, the whole thing can be reduced to arbitrary axioms and undefined terms from which certain deductions are to be performed.

Separated from its base in reality, the question has to come up, why does reality conform to mathematics? This question has come about because induction has been regarded as an imperfection by some people. Euclid might have laughed at what has been taken by some moderns to be a supersession of his opus—the notion that the propositions are mere amplifications of axioms giving meaning to undefined terms.[249] These people have been met by intuitionists who denied the law of the excluded middle, set theorists who debated over whether to inject the so-called axioms of choice and others. The difficulties in set theory has given impetus to—if it did not altogether create—the separation into pure mathematics and applied mathematics, so eloquently lamented by Kline in his *Mathematics: The Loss Of Certainty*.

Before the end of the third decade of the 20th century, this search for deduction faltered before the skeptical argument of Gödel who taught that within any such deductive mathematical system, there are statements which cannot be justified by its axioms; that it can only be justified through some axiom outside of the system. From this, one is led to conclude that if each criteria depended upon a further criteria, one is left with a series of Chinese boxes, terminating in what?

What Gödel calls "undecidable" are really contradictions; what purports to be complete cannot be. The older logic would simply have concluded that there was something invalid, either with Mr. Gödel's argument, or with what he was criticizing. But the devotees of modern mathematics hold it to be but a burden they must carry—that it is an amphibole in human reason. They are in the lineage of Kant in that they hold because of the structure of human reason, there are certain antithetical propositions which cannot be logically resolved.

Gödel avoided one difficulty of set theory. By dealing with recursive formulas only, i.e., with finite sums, he was able to avoid talking about infinite totalities. He considered only whole numbers. He argued that a system like that of Russell & Whitehead's *Principia Mathematica* (which attempted to contain all that could be said about numbers) would have certain undecidable propositions which would prevent it from being complete.

Gödel's position does not apply to any of the important positions taken in the book the reader is studying:

(1) His argument applies to attempts that are founded on some kind of symbolic logic, as he himself admits.[250] What we have shown so far should be enough to convince any reader that the infinitesimal and the number line are not remotely related to that kind of thinking.

(2) A complete account of whole numbers is not even attempted in the pages of this book, only the distinction between numbers which are capable of integral subdivision and those which are not. This distinction is true.

(3) If there were a complete knowledge about whole numbers or any other branch of mathematics, the wider abstractions would embrace the lesser ones and the truths arrived at through deduction would be consistent— neither contradicting each other, nor the wider abstractions. Any axioms induced should be common notions and not aetherial depositories into which "derived" theorems are wrapped with theoretical gossamer. As Samuel Blumenthal, a critic of modern teaching, said so well, *"Mathematics is made to be consistent."*

(4) Gödel was a Platonist.[251] Indeed, the whole movement of which he was a part could be characterized as neo-Kantian Platonism, i.e., it simultaneously accepted self-subsisting ideas and Hume's notion that there was no necessity in the facts presented in sensation. That is not the standpoint of this book. Even if God were to use Platonic ideas, it would not be the proper beginning for human science, since we are derivative creatures and could not put ourselves in His place with the definitude

which these theorists claim. Our basis is sense, reasoning from sense, inborn ideas, and reasoning from them. One of the leaders of the foundational movement was John Von Neumann. The disillusion he eventually felt could hardly be better expressed than they way he did: "As a mathematical discipline travels far from its empirical source, or still more, if it is a second or third generation only indirectly inspired by ideas coming from 'reality,' it is beset with very grave dangers. It becomes more and more purely aetheticizing, more and more purely *l'art pour l'art* . . . . In other words, at a great distance from its empirical source, or after much abstract inbreeding, a mathematical subject is in danger of degeneration."[252]

The number line is not fully deductive. So far, two kinds of numbers are known, rational numbers and irrational magnitudes of differing types. Also certain non-numbers existing below the level of the finite, the positional points, have been identified. The irrational numbers were not discovered, strictly, by deduction. The problem was finding the hypotenuse of a right isosceles triangle. A rational number was shown to be impossible, yet a magnitude incommensurate with the rational was shown to exist. Recognizing the incongruity was reasoning from general to specific (deduction); the realization that what was discordant with the rational but had to exist was an act of discovery. Abstracting the existence of a hitherto undiscovered type of number, the radical—specifically $\sqrt{2}$—involved reasoning from specific to general (induction). An abstraction was drawn; the abstraction was understood to be a universal—an example of a type. It would still be a universal, even if there were only one of a kind. But as we know, they are beyond number.

The author's discovery of the solution to the mystery of $\sqrt{-1}$ took place in a flash. It was too fast to recall the steps involved. This was also the situation with the discovery that the discarded terms of the derivative had to be something.

It is not known what went through the head of the inventor of the number line. The discovery of the infinitesimal required both deduction and induction, as well as the realization of its existence. The full implications of the point must be thought through. The Gödel theorem does not apply to it, for his theorem is based upon the relationship between axioms and their theorems within the context of modern philosophy.

In the final paragraph of his history of set theory, Ferreirós observes: "Today we do not feel so sure that axiomatic set theory is the ultimate foundation, but certainly all mathematicians are accustomed to using it as the basic language to teach and learn in order to become a mathematician."[253]

About thirty years ago, the educational establishment in America decided to base the teaching of arithmetic on set theory. The big idea was that since they believed set theory to be the foundation of mathematics, it had to be the proper place for young people to begin how to calculate. This "New Math" confused so many youngsters that it was soon abandoned. Yet, if it had actually undergirded arithmetic, it might have worked.

The actual foundation of arithmetic is the way that children have been taught for generations. Examples were held up before them. From that they induced such concepts as addition and subtraction. Later on, they were shown more advanced ideas like multiplication and division, etc. Their innate ability to form concepts of quantity and apply them was put to use in the context of numbers. "Practice," as they used to say, "makes perfect."

The foundation of mathematics is very similar to that of learning ordinary words. Indeed, one cannot learn about quantity without knowing something about quality. No literary person would say that "the set of Shakespeare's published plays is subsumed in the set of all that is written in the English language." Rather, they would say that "all Shakespeare's known plays were written in English." Talking the other way is not more exact; we do not need to have a list of everything published in English in order to know whether Shakespeare wrote every one of those plays in English. Similarly, we do not have to know every irrational number that could ever exist in order to define its essence. There is no good reason why the mathematician should use such unnatural language.

The number line affords an easy and natural way of understanding numbers. Learning about infinitesimals allows the older student to understand integration directly and the derivative indirectly.

Logic is more than mere inclusion and exclusion.

CHAPTER XXII

THE STURDY UNIT

The realm of the finite consists of that which is in principle countable or measurable (at least, approximately) by men. Any finite interval can be designated a unit. Being continuous, it includes every single infinitesimal along its length. The key term, 'designated,' shows that it requires a mind to form it. Sticks and stones would still have the same characteristics without someone to measure them, but it takes a mind to call some length a unit.

The number line itself is mental, but that idea is capable of being brought to existence in space, extending from the least to the highest known, even though men are not able to implement it in its totality. On a much, much grander scale, it is similar to an engineering drawing of something that is never built. What counts is its consonance with the facts of reality.

In Chapter VII, it was shown that a theoretical minimum unit cannot be found through subdivision. Since any rational fraction could itself become the basis of a new unit, what would be left after all such fractions had been eliminated would be nothing except irrational magnitudes on the one hand and positional points on the other.

In Chapter III, it was shown that the very basis for distinguishing a unit from a mere string of infinitesimals is the presence of irrational magnitudes. Could one find the theoretically minimum unit by seeking the smallest irrational magnitude? Such an effort would fail for the same reason as the other one did. A number like $\sqrt{2}$ is a length, and that length could itself be chosen as a new unit.

Let us continue this search. If we knew how many infinitesimals there were in a given unit, each of them would be a fractional part of the unit. In other words, if there were n infinitesimals in a given unit, and we knew how many there were, each of them would be $1/n$. But we humans do not and cannot know this. The infinitesimals within it are immeasurable. Therefore, when we operate from the finite, the only level of which we have direct experience, some magnitudes within a unit are not fully explicable to us. We have to approximate their values in terms of what we can know.

The immeasurable is not just relative to our instruments It is a metaphysical limitation, beyond which men cannot go. The infinitesimals are not finite parts of anything; they are locations. The rationals are sub-divisions explicable in terms of the unit, which, of course, is finite. But because the quantity of infinitesimals within the unit is so vast, there are sub-lengths which cannot be stated as exactly so much of those which are measurable. Any non-rational number has a length which does not fit in with any ratio men can find. The common irrationals and the transcendentals can be approximated; the positional magnitudes cannot be. The positional points are even more unreachable.

To know the square root of two, for instance, we would have to know how many infinitesimals it contains, relative to some particular length of unit. In order to use the Pythagorean formula, however, the rational lengths would also have to be stated in terms of the quantity of infinitesimals. The finite and the infinitesimals cannot be used in the same formulation; such mixtures would be incommensurable, which would simply restore the old problem in a different form.

Yet, the idea of a minimum linear interval beckons us onwards. What is it about a line under construction which prevents the next point to be added from being out of orientation, forcing it into a slant? Is there some length, beneath which the line cannot be divided without disintegrating?

Let us attempt a direct construction. Starting from the beginning of a line, or 0, there must be a point above it, and then one above that, and so on. It would seem that a point will eventually be reached at which the border of the level of the infinitesimal will at last be reached. Yet, we are told that there is no theoretical minimum line.

Such an attempt would be vain. The truth is we cannot begin that way, hoping eventually to transcend the level of the infinitesimal, like the change in dimension when a figure turns a corner. The difference has nothing to do with how many or how few steps remain before something analogous to a corner point is reached. Lines are not consciously constructed point by point. Men have no way finding so much as one of them, let alone counting an assemblage. We have to start with an interval. As soon as the unit is established, the points on the interval are present. The points within the interval between 0 and 1 are not free creations; they are not brought in one after another. There is no laying down of 1 point + 2 points + 3 points + 4 points . . . . , with the ellipsis hiding, as it were, behind a veil, the point at which the sub-finite suddenly becomes the finite. The points later to be identified as upper boundaries of subdivisions and incommensurables are present the moment the unit is established; they are there, because the quantity of points in the space taken up by the unit was present before some human decided to create that scale. The continuity of space is pre-existent; the locations were already present. The relationships between these locations that could later be coded as numbers already existed; the same with the irrational magnitudes and the positional points. All that remained for men was to establish the unit and make discoveries within it. First were the locations in endless space. The finite numbers we give them come later and are forever rooted in our inherent finitude. The infinity before and aft of any of these real numbers beyond zero on that line is countless. The line itself is engulfed in countlessness. Yet, the facts we have identified and marked as numbers are part of our knowledge. Our knowledge of $\sqrt{2}$ resulted from a mathematical operation. The interval marked as "2" was not generated as the linear product of that radical times itself. The points establishing its interval already existed when the number system was applied to them. And that is nothing to despise.

Let us try by going to the extreme opposite of the infinitesimal. Consider $\mathfrak{C}$, the symbol for an unlimited quantity: Would not a line composed of an inexhaustible quantity of units contain as its minimum length the smallest unit? And would not this be $1/\mathfrak{C}$?

That would be the wrong answer, for $\mathfrak{C}$ has no end. The complete unit cannot mirror $\mathfrak{C}$; since it cannot do that, it cannot mirror all of the potential rational magnitudes which would be required.

This fraction could not be an infinitesimal, either. If $1/\mathfrak{C}$ were an infinitesimal, what of $1/\mathfrak{U}$ and $1/\mathfrak{I}_\mathfrak{u}$? ($\mathfrak{U}$, it will be recalled, symbolizes the number of infinitesimals in an optimal unit, and $\mathfrak{I}_\mathfrak{u}$ stands for a quantity of linear units equal to that of the infinitesimals within the unit, $\mathfrak{U}$.) The infinitesimal cannot be the reciprocal of all of these at once. The minimum must be on the level of the finite. The infinitesimal is not the endlessly small, but the smallest element of space, the location.

Much of modern mathematics is wrapped up in fallacies of this sort. The Weierstrassian concept of continuity is really an argument for the internally infinite in quantity, i.e., for an $\mathfrak{C}$ enveloped inside the unit. By so defining it, the modernists put the unlimited within the limited, leaving themselves no choice but to argue that an infinity inside another infinity can be equal to the latter. As a result, they mentally evict the point from space, thereby putting it outside reality and converting it into anything the fabulist wishes it to be. Many times, in modern mathematics, points are used in ways that contradict the definitive nature of the infinitesimal. Since the infinitesimal is an absolute, the metaphysics of such people have to be wrong. (For an example, consult a text on projective geometry.)[254]

This is why the ideal number line imagined by Dedekind and others does not and cannot exist. It would have to be endless, whereas such lines have only a potential existence. Moreover, the unit itself would also have to be infinite in length in order to mirror it, an impossibility.

Mention has already been made of the mistake in thinking that the asymptote will go on and on forever, when $\mathfrak{C}$ surpasses its infinitesimals. What is humanly unreachable is not the same as that which cannot be.

Series projecting limits are potential infinities as well, but they do not and cannot exist apart from the number line; rather, they depend on it. The quantity of infinitesimals within a unit is infinite only in the sense of being uncountable, but not in the sense of being limitless; this means that many geometric series projecting limits are approximate in a double sense. Not only do they fail to hit the mark, but the unit with which the mathematician is working may have to expand as he proceeds—as explained below. In the case where the mirror of $\mathfrak{C}$ is required, this would be impossible. (This does not take away from their value in exhibiting significant patterns.)

Since the two extremes are outside human competence, let resume our search for the minimum linear interval by starting from the middle.

Consider two lengths used as units, one longer than the other. The larger length must contain more positions. Where do the extra points come from? Suppose first that there is no difference in the quantity of potential rational and irrational magnitudes between the two units. If that is the case, it could come only from the positional points.

Assuming this to be true, there should be a way of identifying the extra points in the longer unit. But such a way does not exist. We simply cannot find the positional points; all that is known is that they have to exist. How many there are in a unit of some specific length is fundamentally unknowable.

But it is not correct to think that the difference in length is due to positional points. Actually, on a straight line, the positional points would be the same, regardless of the length of a unit, whether it be shorter than a micron, or extend beyond a mile. This is because any point elsewhere within the unit would be at some finite interval from the beginning. We shall see that it depends upon rational and irrational magnitudes alone.

A prime number is one which cannot contain any factors (or as some would say, other than 1). Consider the highest prime known at this time, a number with over a million digits; this is not an arbitrary figure just thrown out, but has been established through calculation, using advanced computers. Let us call this highest prime number known, "P." In order for a unit to be completely consonant with modern knowledge, it most likely would have to contain the fraction, 1/P. Just so that there is no misunderstanding, this is not 1/P of P, which would be 1; this is the Pth part of a unit. A unit which did not contain this fraction would be shorter than one that did. Units, therefore, differ with respect to the quantity of rational and common irrational magnitudes possible to them.

The length of an actual unit is continuous, since nothing can be inserted between its parts. As men work with numbers and extend their range, the unit which they work with must grow in consonance with the requirements of the user. This means that the operator is actually working with a unit that can change in accord with his or her needs. The unit can be made as long as required.

This tells us something about the maximal size of a unit relative to the present state of human knowledge. Does it tell us anything about the minimal unit on the number line?

Let us consider the related question as to how the unit grows. Just as the addition of a higher prime number extends the length of the number line, so the unit itself is thereby increased in size. This is because the unit must include its reciprocal,1/P, because this same number also occurs in the fraction, although only in the denominator.[255]

The introduction of a new rational number brings with it new roots and new irrational magnitudes. This could also be obtained by simply increasing the number line. Although a non-prime N would be a composed into factors already present, the reciprocal of their product would have to be a fractional part of the unit. Its factors cannot substitute for it.

Except for the theoretical minimum, the smallest rational quantity is found as the reciprocal of the largest known integral number, itself a potential infinity. Furthermore, this smallest rational fraction does not shorten the unit, but rather lengthens it.

The number line, therefore, is not to be a one-size-fits-all situation. The size is determined by the largest whole number known. This number must be reached by calculation. Mere dreaming is not enough, since the unit, once it is defined, puts into order the points, the relationships of which establish the possible rational subdivisions and the incommensurables as well. The reciprocal of this highest number known sets the size of unit.

Without this sliding scale, adjusting to discoveries of new natural numbers, the unit would not be on a line; in fact, it would be out of space. Continuity exists because infinitesimals follow one after another on a line without a break. Instead of arbitrarily introducing new numbers in the midst of a fixed interval, as the Weierstrassians would have it, the unit itself is increased through the introduction of the smallest rational fraction. The continuity is supplied by the infinitesimals within the number line, which, being mere positions, do not change. What changes is their evaluation. They must necessarily be recalculated as a result of the increase in the unit, which in turn is caused by the advance of the whole numbers.

Stated differently: $\mathfrak{U}$ and $\mathfrak{I}_\mathfrak{u}$ both grow, but $\mathfrak{U}$ brings about the growth of $\mathfrak{I}_\mathfrak{u}$, as does the latter, the former. Thus the number line, although a ∞, grows both from its extension of N and from the expansion of its parts. That being the case, there cannot be a Platonic unit out of which actual lines are but imperfect realizations. It would not be anchored in reality. There is not and cannot be a $1/\mathfrak{C}$.

Once more, the element of time is present in mathematics, showing it to be a fallacy to attempt to exclude it from that science. Motion also is present, for the advancement of the number line is a process of both invention and discovery.

But this does not eradicate the problem of the minimum unit. Consider the case of dividing a line by halves. Within any line, there is a definite quantity of infinitesimals. The process of division cannot go on forever; sooner or later, it must exhaust the quotient. At some point, it must stop. There is nothing left to divide.

Let us examine the situation near the end of the process. It would not be the case that the rational divisions would cease, leaving only the non-rational magnitudes. That would be impossible, since the latter depend upon the rational numbers for their intelligibility. What would be left is a string of infinitesimals. Were it not for the fact that they would be too small to detect, one could count them. Unbeknownst to us, the unit would be replaced by a sequence.

This suggests that there might be a minimum linear unit. Yet, finding it has eluded analysis. Beginning at the beginning with infinitesimals did not work; neither did considering endless infinity. Searching amid the types of magnitudes did not either.

If there were a way out of this impasse, it would lie in understanding that at some point in the divisions into smaller and smaller units, there comes an interval so tiny that if it were but one infinitesimal shorter, it would contain no irrational subdivisions. That would be the minimum unit. Below that minimum, the infinitesimals within the unit would be countable in theory.

But this is a distinction with a barely discernable difference, for infinitesimals are too small to detect anyway. The quantity above which lie the incalculable and below which lay the theoretically countable is humanly unknowable.

What is known is that the remnant of what was the minimum unit would be stable throughout its transition from a mere sequence of infinitesimals to its final exhaustion. From the human standpoint, it would be the same throughout.

In the fifth chapter, it was shown that a unit cannot begin with fractions or radicals. Before there can be any rational or irrational magnitudes along the length, there must first be the positional points. These points belong to the unit itself. As the process of division nears its end, the remaining interval does not crumble. Far from being something alien added on, these points were there from the beginning.

This shows that the linear interval is stable, that a straight line can be folded until there is nothing left. Needless to say, there are practical difficulties. If a person attempted to do this with a material unit—the only kind known to exist— he would soon exceed both his reach and his grasp. A mathematician has argued that if a person were to fold a single 8-1/2" x 11" sheet of common bond paper 51 times, its thickness would reach past the sun.[256]

In order to be able to stop at the last infinitesimal, the same divisor could not be used throughout. Otherwise, the mark would be missed. Furthermore, a person would not know which divisor to use unless he knew in advance not only the final position, but how to count off all of the positions which preceded it—a human impossibility.

The reader will recall how in Chapter VIII, the last two digits of .999 . . . could not be 9 and 9. They would have to be 9 and 1, or better, $9/10^{n-1} + 1/10^{n-1}$. The last term has the same denominator as its predecessor. The process of division is succeeded by addition.

That is how it is with a rational number; granting that the underlying unit is of suitable length, we can know what the last two positions must be, even though we cannot know the value of "n", for to know that we would have to be able to count the positions which preceded them. Even this, we cannot do with a common irrational number. With mere positions along a line, the attempt is so far beyond human competence, only one more thing can be said about it; what we have said already, namely that the realm of the non-finite can be touched, but not comprehended.

Men are unable to provide themselves a way of discerning a unit so short that anything shorter would be immeasurable.

Now the question left unanswered in Chapter XIX on the integral can be answered, namely, what happens in the integral when the curve is nearing the zero point on the base axis (X) below the level of the finite? If there are no vertical lines from base to curve in this portion of the area, how can an accurate determination be made? Wouldn't this be an approximate for the tangent theory also, a case of a theory not quite meeting the requirements of exactitude?

The answer to this is as follows: This tiny part of the area is as densely packed with positions as the rest of it. There still exist the remaining points along the base and corresponding to them, the remaining points of the curve above. Each of the points on the curve has a tangent, even those near zero. Between any basal point and any corresponding point on the curve, there are a string of infinitesimals, not enough to constitute a distance, which is a finite determination, but a specific quantity of them. These positions both separate and connect the points on the line to those on the base. Although they are not lines, they are components of what would have been lines if the finite level had been reached. They are, so to speak, "proto-lines." Together, they fill out that insignificant corner in an orderly way. With them in place, the curve can be calculated from 0 to any point capable of being discriminated in between. There is no hiatus anywhere.

Now, let us resume our discussion of the smallest interval. Consider a circle, where the points along the perimeter are in line with the circle's center, but not with each other, since the tangent is different at every point on the rim.

In Chapter VII, it was shown that a minimum curve cannot be produced through division. Is there a size beyond which the circle collapses? Or is it stable, like the straight line?

At first thought, one might think that it is like the straight line. As its radius decreases, the quantity of points on the perimeter would also shrink. When the radius was but one infinitesimal in sub-finite thickness, there would be a minimum circle around it. Were it not for its size, the number of points in the perimeter would have been countable. Once this radius of one infinitesimal is removed, there is no more nimbus of points about it.

This cannot be. Without a finite radial distance, let alone a line, there can be no π. This pi is an irrational number. In order for π to exist at the sub-finite level, there would have to be irrational quantities beneath the level of the infinitesimal, a thesis which was refuted in Chapter VII. Neither can it coalesce into a single point, for there is no shape to an infinitesimal.

The circle is not stable; there is a shortness at which it disintegrates. But it does not shatter; to the contrary, it dissolves at once. For to shatter, there would have to be remnants of a unit, which would then become units themselves. It can only disintegrate into infinitesimals. When the quantity of magnitudes get too small, the unit disappears. Abel noticed something like this when he discovered that the balance of rational and irrational coefficients broke down at the fifth degree.[257]

It follows that the minimum circle cannot have a radius of one infinitesimal. Yet, π is a definite position on the number line. The minimum circle would have to contain enough infinitesimals to reach π—an exact quantity in excess of the position obtained by three times the length of its radius. Obviously, a radius of one or a few more infinitesimals would not suffice. To support that existence, the minimum radius would have to be finite, i.e., the quantity of infinitesimals within it would have to be immeasurable. (The same with the sides of the minimum angle).

In short, the minimum circle has on its perimeter the least multiple of the quantity of infinitesimals that are required to reach π.

This would have to be the case with other curves as well, for they also are require π. Once more, we see that curves cannot be reduced to straight lines, and *vice versa*.

But no figure, whether straight or curved, is simply infinitesimals. They cannot act. They remain what they are, regardless of whether or not a line has been drawn through them. Since there is nothing between two neighboring points, there can be no hooks in them which support a line, the failure of which could lead to the latter's disintegration. They constitute the space that would be filled by a figure.

Studying the infinitesimals enables us to understand what is possible in space. To be actualized, however, some means must be provided. For man, that means is material. The thick lines of a good construction made by a material substance take up space, i.e., occupy infinitesimals; within these material constructions are the possibility of geometric lines. Michelangelo grasped a similar fact when he famously said that within a block of marble there existed a certain statue.

It has been established that the existential line is a sliding scale set by the highest whole number known. And this sets the size of the spacial unit through its reciprocal. But there is a theoretically minimum unit. It would be the length in which there would be no intervals below that reciprocal, only positional points. This would not be a Platonic idea, but an actual possibility in space, even though men could not know it.

Yet, when it comes to minimal length, the limits are set by the inherent capacities of that out of which these lines are made. To the best of our knowledge, energy is the genus, of which light and mass are the species. If this is a fact, no geometric line can be shorter than the minimal capacity of nature to effect this.

What is light? At present, there is dispute as to whether light is a particle, or a wave, or both. We shall not enter into that here. But, whatever it is, there must be a minimum size, and it is this that will determine at what point light will break down. With regard to the atom, this writer knows of two schools, the planetary model and the ring model. Whatever is the minimal thickness of objects possessing mass, this sets the practical minimum size for a material line. This is a worthwhile goal for physical science. (Beyond that is what God can do, a subject outside of this inquiry).

Still, there is a cost when dealing with actual material lines, rather than possible lines in space. Sir Isaac Newton was aware of this: "But because the parts of space cannot be seen, or distinguished from one another by our senses, therefore in their stead we use sensible measures of them . . . . And so, instead of absolute places . . . we use relative ones; and that without any inconvenience in common affairs; but in philosophical disquisitions, we ought to abstract from the senses, and consider things themselves, distinct from what are only sensible measures of them."[258]

He was almost right. Failure to understand that both can exist in space might lead to problems on the practical level; "data" arrived at from experiments set up by men ignorant of this fact can lead to false conclusions. If such a theoretical mistake were in turn accepted by some engineers working on an exotic machine, the enterprise might fail.

> "For want of a nail, the shoe was lost,
> For want of the shoe, the horse was lost,
> For want of the horse, the rider was lost,
> For want of the rider, the battle was lost,
> For want of the battle, the kingdom was lost,
> And all for the want of a horseshoe nail!"[259]

Is there an optimal length for man living on earth? Since it is the case that all measures are not completely inter-translatable, the possibility immediately opens up that there might be one entirely consonant with our nature and our environment.

CHAPTER XXIII

THE INFINITESIMAL OF TIME

Time is not a mere human construct; sun and tide might have existed, even if there were no men or animals. Neither is time material; it is not a clock in the skies clicking off the eons. Time is a fundamental fact of reality.

Some think it is only another name for change. But, as I wrote in an earlier book, *The Stance Of Atlas*: ". . . all examples of change are not temporal. When we look at the corner of a house and notice a change of direction, we are not observing a temporal process . . . . In a circle, there is a change of direction at every point. These are *spatial* changes.

"Time does not always involve change even in situations. Part of the definition of the law of contradiction in its common form is that two or more existents cannot exist at the same time and in the same space. Here, the absence of temporal change is emphasized.

". . . . Temporal events are not the only phenomena which exhibit successions of states. In a line segment, there is succession. Yet, herein is not an idea of time, only of space . . . . . A succession of numbers—1, 2, 3, 4, etc.—is just that, a succession of numbers, but it is not a temporal one. We can say that the number '3' comes before '4', but there we mean that '4' is '1' greater than '3' or we simply indicate that '3' is to the left of '4'—not that the number three is *earlier* than the number four . . . . Time cannot be understood in non-temporal terms. We sometimes speak of a "before" and "after" in space, but this is more accurately described as 'in-front-of' and 'in-back-of.' Coexistence in space would describe a condition of simultaneity, but they are obviously not identical."[260]

The infinitesimal of time is called the "instant." It symbol is $\bullet_t$. *Just as there is no interval within a spacial point, so there is no multiplicity within an instant.* And that is where the similarity ends. The instants succeed each other, but the predecessor ceases to exist when its successor is born—without any overlap. There can be no "time" between any instant and its successor. There is only the eternality of now! now! now! without a break.

As the infinitesimals of space are not elements, but simply locations, the instants of time are elements. But they are not discrete in the way defined by Edward Huntington. The instant after they come into existence, they are no more. Yet, there is no becoming, no temporal gap between the instant that expired and the one which succeeded it. Neither is there a suture. The second replaces the first without an interval. There is no beat to time, no rhythm. Since this eternal replacement is without breach, it is also continuous.

Time is like a point; unlike space, however, it can have no dimensions.

In an earlier book, this was graphically represented. "Fact" is the ultimate category. It is more fundamental than existence. The one thing all facts have in common is that they either are, were, will be, or might be. If something does not any longer exist, then it is a *fact* that it can only be classified under non-existence.[261]

FACT

Non-Existence   Existence

The one thing all facts have in common is that they either are, were, will be, or might be. Where free will exists, the future is not fully determined.

Since this work is not about the various types of existence, the drawing showing the major categories under "Existence" will be skipped. Needless to say, both "Time" and "Space" are among them. This time is the instantaneous present discussed above.

Now, consider the category of "Non-Existence." [262] Each instant of time exists, only to cease existing and be replaced by another.

Non-Existence

Past   Future

This means that the past does not exist. Neither does the future. Unlike space where every point is simultaneously present, past instants do not co-exist with the present one. Space always exists, but only the present instant does.[263]

A temporal interval is a duration, with a beginning, a middle, and an end. Neither a duration's beginning, its middle, nor its end (usually)[264] exists as such. It is just a concept about organizing that which is no longer. Even though it might be factually sound, it is never existential.

To represent duration, it is common to imagine a line wherein past events of importance are labeled milestones. This representation of a "length" of time in spacial terms is invaluable, even though it is done through a different kind of fact. These lines are usually drawn from left to right like ➔, but they could just as logically be represented from right to left: ←. There is no such thing as time reversal. The instants simply are and then are not. The past cannot be understood fully in non-temporal terms.

The very thought of representing the past in terms of present instants is unthinkable. Since these instants no longer exist, they cannot be present. The alternative is to borrow the line from space and mark the events as if they were spacial points.

This is paradoxical. Although the past does not exist, it can be objectively represented. In this book, there are references to previous centuries, even quotations from people which have passed away; the attentive reader is thinking thoughts which were presented earlier. Given the concept of duration, we can also subtract time. If, for instance, today is Sunday, then two days ago, it was Friday. A man who was born in 1938 would be about 66 years of age now (2005), and so on.

Contrariwise, we also use temporal terms to indicate a crucial attribute of space, its com-presence, the fact that it is all in existence at once. We think of the third leg of a triangle existing at the same time as the other two, or even of the universal simultaneity of the points. Because of this attribute, we mean something quite different when we say "A succeeds or follows B" in a spacial figure or when we say "A follows or succeeds B" in time. But the difference between the two is that while we must use the non-temporal to indicate duration, we can work with spacial figures without contemplating the simultaneity of their parts.

(Some readers will want to ask why error and illusion were not included as the third category under "Non-Existence." The short answer is that the psychological fact of having a false idea is real, but what that idea itself purports to be—a fact—is not. This was discussed in some detail in the earlier book, most of which is reproduced in the endnote.)[265]

Actual time is the present, or $\bullet_t$. This appears to conflict with our everyday notion that it has a triune structure—past, present, and future. but the conflict is only apparent. This concept includes the two categories of non-existence, that which once was, but no longer is and that which will be or might be. But it is only the one in the middle that actually exists.

Although non-existent, the past is objective. Colloquially stated, duration is the passage of time and the instants that have passed away were once real. But this is not an exact statement, since past instants do not go away to some other place. They simply are not. Because space is present everywhere at once, it affords the means to order human experience. Our language has difficulty expressing this thought, yet we all recognize its truth. This difficulty arises from the fact that all human consciousness is of difference,[266] but what has been changed no longer is what it was. Space, then, (which does not change) is the background against which the triune concept is brought together and comprehended.

The future, although it does not exist, is ineradicatably implied in potential infinity. It is a future which never arrives. It was thought by many 20th century mathematicians that their work would be better if it were purified of non-static elements. But even with the Weirstrassian definition, there lurks about it a suggestion of the temporal. To say that f(x) is continuous at x through x_0 if given any positive number ε, there exists a δ such that for all x in the interval $|x-x_0| < \delta$, $|f(x)-f(x_0)| < \varepsilon$ is to suggest an existence at a certain moment. If a number δ is greater than some x interval, there must *simultaneously* exist a number ε that is greater than the f(x) interval. A fog consisting of brackets cannot not blank out that fact.

Actual infinity is best represented by space. Even an infinite unbroken sequence of time must be shown in spacial terms. Yet, time is as continuous as space. Just as the locations of space can only abstractly be spoken of as the elements of its continuity, this is also the case with the instants, but in a different way: they can never be co-terminal. Instead of a next-to, there is a just-after.

The independence of time is well understood by health professionals. As a physician recently put it: "Old age is not old age . . . . It is your body failing to repair itself."[267] A man like Jack Lalane who could do more at 70 than most men could do at 40 was not any younger than they; what he had done was beat the averages. And an average, although it may be truthful, is a non-fundamental abstraction.

But the fact that time is neither mental nor material does not mean that it is without properties. In that regard, it is like empty space. If the physical universe were at rest and there were no motion, at each instant there would still be a before and an after. If there were not even motionless matter, only a blank concavity, there would still be the constant and continuous replacement of instants.

The instant is not a zone in which layers of before and after are constantly pealed off in an infinite process, leaving no center. Were that the case, there would be no present—a contradiction. Neither the prior nor the subsequent are intelligible in absence of the present.

Since there is no division within the absolute instant, there can be no phase difference between one instant and another. No instant can start on one planet while the instant on the other planet is part way through. Any instant has to be the same everywhere.

The same instant exists simultaneously throughout space[268]; and a point in space is the same at every instant, regardless of whether a material object is occupying it or not. In summary, there is no time which is not a time and there is no place that is not a place.

This does not contradict the belief most physicists have about the measurement of time being relative. Men cannot perceive the instant. The time is past before it is registered on consciousness. Any measurement of time invariably uses material means which exhibit some regularity and is relative to the motion of the instrument being used. A physicist may come up with the conclusion that an observer cannot detect whether two events far apart in outer space had happened simultaneously. A few would carry that case so far as to assert that two clocks standing next two each other means two different reference systems which can never be completely synchronized, but even if that extreme supposition were to prevail among physicists, it would have no bearing on the existence of the absolute instant. Whether these clocks can finally be correlated is a question that is best left to those who attend to such things—the physicists. But even if they are never able to do this, the notion of simultaneity cannot be expunged. The act of observing an event and the observation of the event must be at the same time. It is impossible to change this. The person who thinks he can is welcome to try.

Without presupposing simultaneity on some level, one could not compare the measurement of an event as seen by observer A with that seen by observer B. For, if there were never any temporal coincidence anywhere, no comparison like that could be obtained. For example, Thucydides observed a battle between the ancient Athenians and Spartans and some other observer sees a battle waged by Napoleon which had some similarities. Without the concept of simultaneity, a measure of their "distance" in time could never be made.

The deniers end up with two atoms spinning next to each other, each with its own "temporal" perspective. But how much would this unsolvable difference be between the t of one clock and the t' of the one next to it ? Not very much; the interval which this theory supposes can never be closed would be tiny. The closer the clocks, the more insignificant the difference. The discrepancy does not come from some metaphysical difference between the two times, but because of physical problems. But the instant is really the same everywhere. While the tick of one of these particles might be indistinguishable from the tock of the other, the instant is sure. It is omnipresent, regardless of whether these relativists are right in their assumption that they can never be observed by men.

The finite divisions of time are extrapolations from finite divisions of space. Duration is thought of in terms of a line. But it is not. Unlike any line in space, it does not exist. Like any spatial line, however, it is finite. This provides no problem for the future; no matter how many years or eons pass, it is finite. But when one considers the past there are some interesting conclusions. Starting from the present, no matter how many finite intervals one projects to the past, the furthest removed from this moment must be finite.

This observation does present some problems. Our projection of the past must be irretrievably finite. The line backward is like the lines drawn in space; they are only potentially infinite. What existed once in the past is not now. Unlike lines in space, the temporal ones we draw in our mind to indicate the past have no actual correlate. These are analogical devices of consciousness; the shadows on a movie screen have more reality. Time is not a line; it is more like a point, a replacement of that which is destroyed.

This should not be taken to mean that history is not true, either; Julius Caesar might have existed pretty much as he was depicted by Plutarch,

Suetonius, and others. The issue is metaphysical and goes quite beyond the power of man's power of reductionism. It is well understood in the Bible. In the 22nd chapter of Matthew, there is the story of the Sadducees, men who did not believe in the Resurrection, coming to Jesus: He answered: "But concerning the resurrection of the dead, have you not read what was spoken to you by God, saying, I am the God of Abraham, the God of Isaac, and the God of Jacob. God is not the God of the dead, but of the living." God is not the God of that which does not exist. Consistently acknowledging this fact, Jesus preferred to speak of those whom He was about to raise from the dead as *asleep*.

But does the possibility of *division* by zero in abstract time also exist? Indeed, it does. Take for an example, 60 seconds/$\bullet_t$, where $\bullet_t$ stands for the indivisible instant. The duration of sixty seconds in comparison with the instantaneous beginning of a duration is infinite in the sense of being beyond human possibilities of measure. The same with the duration of 1 standard second in comparison with the instant. Within that temporal interval, an immeasurable infinity of instants have fallen. Whenever that 0 is the beginning of a finite temporal interval or duration, such <u>division</u> is possible.

Consider the concept, eternity, symbolized here as $\mathfrak{E}_t$, the Old English "E", which stands for endlessness, but with a subscript "t. This does not refer to time, *per se*, but to something which persists, to that which endures without end, which does not stop at some instant. An early 20th century translation of the Bible refers to the Old Testament Jehovah, as "The Everliving."[269]

Why does time seem so much stranger than space, although it is also omnipresent?

Knowledge of time's existence does not come from the senses. This is why time seems so much stranger than space, although it too is omnipresent.

It is because of the wonderful gift of memory that we are able to obtain the idea of time—first that there is a past, which implies a present, which in turn implies a future. It is not an abstraction from sense data either, for a valid abstraction cannot have within it anything which does not belong to it. The idea of time rests upon an innate idea.

We have no way of perceiving an instant of time; the duration of time, we have shown, is a re-construction. It is from memory that we become aware of the existence of the fact of time. Memory is an ultimate power of mind. It cannot be reduced to a set of outer perceptions; there are no tags attached to the data coming from the external world stipulating that it is something present which will soon be converted into a past event. The designation of an event as having taken place in the past is an internal apprehension. Something was added to the stored experience, this being the cognizance of the fact that it was stored—a recognition that cannot be further reduced to something more primitive without canceling the sense of it.

John Stuart Mill was one of the relentlessly empirical philosophers of all time. He attempted to analyze every mental fact into simpler phenomena derived ultimately from sensation and the laws of association. But even he understood that this was impossible to do with respect to memory. He recognized that this ability to retain and recall ideas must be classified as an ultimate power, outside of which no other account can be given. Any contrivance for testing memory's veracity eventually presupposes the very thing to be explained. In other words, that remembrances of states of consciousness-past exist. He knew that the mind has to be more than a series of coexistent and successive sensations, as extreme empiricism would have it. It is aware of the fact that it is a series. Memory should always be distinguished from Imagination. My recollection of having read some lines from Kline's *Mathematics: The Loss of Certainty* yesterday is altogether different from the mental gymnastics performed when I put together the following complex ideas: myself—having read from that book—and the day-before today.[270]

This impression that the event was in the past is undeniable and does not come from the external world; nowhere in that world has anyone found a percept called "pastness." It is conceptual in nature and is implanted within man. It may be that there must first be a sensory awareness of the external world in order to start its action, just as it may require the turning of the ignition key before a powerful automobile can start, but once the necessary amount of experience has been had, nothing contained in the objects being perceived brought about the idea of the past.

The fact of recollection immediately makes us conscious of TIME. The comparison of what has been disclosed by memory and what is being experienced in the external world gives the notion of the present. With those two in hand, recognition that there is a future springs into awareness. This awareness is implicit in the disclosure of the past.

Recollection cannot be reduced to mere retention. A medical capsule might ration the release of the pharmaceuticals within it. But the capsule has no memory. Even within the human body, certain chemicals are held back while others are released. Yet, we are not aware of this happening. Memory involves an awareness of the past, of experiences that are no longer there.

The phenomenon of reminiscing is partly independent of sensation from the external world. Put differently: the fact of time is not recognized first by noting just any change in the external world; no comparison of any two states of some external phenomenon could have originated the idea of time without any recourse to the faculty of memory. Time is not a synonym for the relationship that exists when an object changes its position. It is true that time is used in the calculation of velocity; but so is displacement or distance. Then too, change of position is also used in the calculation of time, whether it be based upon observations of the stars or of the hands of a wrist watch.

In the words of John Stuart Mill: "The difference between Expectation and mere imagination, as well as between Memory and Imagination, consists in the presence or absence of Belief; and though this is no explanation of either phenomenon, it brings us back to one and the same real problem . . . the difference between knowing something as a Reality, and as a mere Thought . . . ."[271]

(Note that in addition to Memory, the great empiricist discovered that Imagination and Expectation were also ultimate and not further analyzable. These other two, we will not discuss in this place, although they do figure in the problems of induction and universals.)

Cognition of time does not arise exclusively from our experience of the external world. The faculty which allows us to recognize that something had happened in the past is innate—the past is an *innate idea*.[272] But this does not mean that we need to accept the notion of Immanuel Kant that time is simply a subjective form in which the mind organizes its contents—that transcendental reality outside of that disclosed by the senses might, for all we know, be *timeless*. That would be a non-sequitur. The faculty of memory is a necessity for understanding existence. When we remember something—say attending a concert featuring Tchaikovsky's 6th Symphony—we apprehend the fact that it took place in the past. The concert is no more. We may not even be able to hum the famous opening melody. We know that we have heard it; and when we hear it again, it will seem familiar. The memory is an apprehension of the concert, although not of the same kind as the original experience; this requires an awareness of the past, the primordial apprehension of time. As John Cook Wilson put it, almost correctly: "Time is not seen, felt, heard, or experienced in general, in this sense of experience; it is apprehended along with the apprehension called experience, and as apprehended it is an apprehension of particular time."[273]

The infinitesimal of time is very different. Its name is the "instant." Unlike spacial continuity, instants exist and can only exist when no other does. Whereas each point in space is surrounded by others, all of them existing simultaneously, the instants succeed each other serially. Like a spacial point, each instant is unique; but, unlike it, the instant that expires is followed by its successor. No two instants can be simultaneous. In order for one to exist, the other must expire. Yet, it is also discrete; the instants succeed each other without a breach. There is no time in which there is no time.

Descartes understood this characteristic of it. However, he thought that at each instant the world ceased to exist, but was created anew by God.[274] His mistake was in not comprehending that time is a fact independent of matter and space. It is important to realize that the replacement occurs simultaneously with the extinction of the old—something which cannot occur in matter or in the change from one concept to another.

It must be emphasized that the past does not exist any longer. Only the present exists. There is no actual duration, only a now! now! now! But we do possess the faculty of memory, a marvelous faculty which cannot be explained as mere retention, for the past is gone forever. Therefore, we are able to concern ourselves with what is not now but once was.

Duration is an objectively understood fact, although most of its constituents no longer exist. It is represented by spacial lines. Like the number line, time is both discrete and continuous. But the number line could never be represented, temporally. For what is no longer, is not. But through the idea of duration, the number line can make sense of the now sequence. The spacial infinitesimal is more easily understood.

One last question for the reader to contemplate: We now know that Time is not material and is, therefore, not an image of anything in nature. From what dead, unconscious process are we supposed to think that Memory evolved? Not from mere retention, for that still leaves the question, Where did the idea of the Past come from when all that we can perceive are presentations of sense?

CHAPTER XXIV

THE INFINITESIMAL OF MOTION

The reader will recall how in Chapter X, it was shown that a stationary machine, such as a giant ball with direction indicators attached to every radius, cannot point in all directions. Yet, it is also a fact that if that object were rotated, it would point to every possible spacial direction, as do the sun's rays. What this illustrates is the fact that motion cannot be reduced to rest, that all is not relative.

Neither space nor time move. Lines have to be made in order to exist; infinitesimals which might become elements of lines are only possible lines until they are drawn. Some figures are generated by more than one kind of motion; the Archimedean spiral, for instance, is a compound of two motions: a line which revolves about a fixed point and a point which moves along the line while it is revolving.

Motion is most easily defined as change in position. Matter moves. Scientists debate as to whether matter ever stands still. Even those who like to think that all is relative have to realize that something is moving. Furthermore, those who think it is like that—namely that every motion must be referred to some other motion—really do believe in a definite state of affairs: they trust in an endless series of references within references, an actual $\mathfrak{C}$. But let us proceed.

Contrary to common belief, speed is not really distance divided by time. In the formula, s = d/t, we *measure* speed as distance traversed divided by the time. But that is not speed itself, but only the means by which it is measured. We do not measure speed by itself. If it were, we could say that distance was so many units of speed times some duration, which would be surrealistic, even for a 20th century physicist. We do not measure time either directly, but through changes in motion shown to be regular, whether it be the fall of sand in a specially shaped glass, a pendulum, a tuning fork, or vibrations in crystal— and even then, it is not the actual instance, but an interpretation of duration that is obtained. The only one of the three which we measure through itself is space. There is a number line, but no number time or number motion. Yet, even with space, as was mentioned earlier, sometimes images of time are used to describe aspects of it which cannot be well expressed in purely spacial terms. And sometimes we say "one place follows another," although they are not themselves in motion.

Neither is motion an interaction between space and time, for they do not interact; to think otherwise is to contradict oneself, for interaction is a type of motion. Because the fact of motion is so fundamental, it cannot be characterized, except by synonyms like "traverse," or by words which presuppose its possibility even where it is not, like "remain."

The path of a moving object must cover every point along its trajectory. It is, therefore continuous, even though the object in motion might not be.

We know that there is motion on the finite level. This must also be the case on the level of the sub-finite.

One can speak of a certain number of infinitesimals being crossed in a single instant. This is of the type introduced on page 157, where $\triangle y$ and $\triangle x$ are both sub-finite; only numbers of identity apply—never those of unity.

Some might find this questionable, asking how can there be motion in an instant? Would this not require the division of the instant, since an earlier part of an action would be followed by a later part in the same instant? The answer is that the motion is not between the instants, but at the instant. The whole of the change takes place at that unbroken suddenness. The action is instantaneous. There is nothing to be averaged out. This is another reason why the rectangular approach to the integral is so misleading; it is based upon the idea of an average among tiny changes. But if the change is instantaneous, no average is possible at the theoretical level.

Moreover, not only is there no contradiction in a speed or velocity of $2\bullet/\bullet_t$, but the continuity of motion requires it. In that instant, two infinitesimals are crossed. Think of it another way: if the instant were divisible, there would be no such thing as a difference of speed at that level; each of the two infinitesimals being crossed would have its own instant. And if there were none at the infinitesimal level, how could there be any difference of speed at the finite level?

There is a minimum theoretical speed. That would be an absolute speed of one infinitesimal per instant $(1\bullet/\bullet_t)$. Nothing less is possible. An expression like $1\bullet/.5\bullet_t$ would have to be wrong; it would not be just another way of saying: $2\bullet/\bullet_t$. This is because there is nothing smaller than an infinitesimal.

There is also a minimum acceleration. It is like the minimum speed. With relative speeds, it is possible to treat an acceleration as an average and have a figure like $(0 +v)/2$ for an object that started from a speed of relative zero. But since non-motion is not a state of motion, such a formulation cannot be made. There is a jump. The symbol for absolute acceleration is: $k(1\bullet/\bullet_t - \bullet_t)$. The dash in the denominator indicates that the two symbols linked are to be read as "instant during each instant." The instant, being an infinitesimal, lacks magnitude and therefore should not be given the same form as duration. The "k" must be a whole number, for reasons already given. It is no more than an instantaneous change of velocity.

The two levels—the finite and the sub-finite—are not just parallel; rather, they are expressions of the same causes on different levels. It is not some pre-established harmony in the manner of Leibniz, but the synchronization of cause and effect throughout.

The mathematics of whole numbers and fractions apply at both levels: a speed of two space infinitesimals per instant must be twice as great as one infinitesimal per instant.

Finite speeds and velocities must bear a proportion to the underlying infinitesimals. Whether it is simple or not is not known. The immeasurable and the measurable are too different. Considering known finite speeds, such as that of light and sound; since these differ greatly, the number of infinitesimals crossed in an instant must differ.

Is a mixed ratio of the finite and the infinitesimal possible? Consider first a sub-finite $\triangle y$ and a finite $\triangle x$: A speed or velocity of $k\bullet/t$, where t is some finite *duration* is at least thinkable. It would mean that the object would rest for a time at each location. To the senses, and indeed, to even the most refined sort of electron microscope, it would appear completely to be at rest. Beneath the level of the discernable, there would be a type of syncopated motion. Its derivative would be the number of infinitesimals traversed in the numerator divided by the square of the unit duration.

What about the inverse? That would be a finite $\triangle y$ and a sub-finite $\triangle x$. Can there be a speed or velocity of $6/\bullet_t$? It would mean that it has had to pass six linear units in that instant. It might be objected that in order to reach the sixth unit, it would have to pass one, two, three, four, and five units before it could reach the sixth one, which means that it would have to pass these lower marks in less time— which would contradict the nature of an instant, which can have no parts.

The answer is that it would not pass them sequentially, but all at once. Such a speed is so far beyond that of light as to be unimaginable. Nonetheless, if it can pass as much as two infinitesimals in an instant, it can do this, providing that the propellant can exist. As Irving Adler once put it, "When there is no change in position *accompanying a change in time*, there is no motion. However, there is no change of time during an instant."[275]

Can there be a rate of change of motion, like $6/\bullet_t - \bullet_t$? This expression is reasonable, providing that it is not taken to be an actual squaring of the time infinitesimal, which would be impossible, since no two instants can co-exist. It would, however, be extremely rapid.

Could there be a velocity like $Æ/\bullet_t$? This means that it would go forever in an instant. This does not mean crossing a finite room in an instant, which we have been considering earlier. It could not pass a finite distance, such as a trillion miles first before it went it went beyond the finite. It would have to be everywhere at once. Such a situation is impossible, because it would be no velocity at all.

There must also be a distinction between curvilinear and linear speed or velocity even at the infinitesimal level. But one must be careful. Certain received ideas must be rejected, such as the notion, expressed by Edwin Wilson, that "the displacement of any rigid body in a plane may be regarded at any instant as a rotation through an infinitesimal angle about some point unless the body is moving parallel to itself."[276] Wilson supposes that in an infinitesimal instant of time, the translation of a body can be conceived as if it were a rotation. But, once again, curved motion is not the same as non-curved motion. A difference still exists between 0 and → 0. The path is continuous, and has the same effect as a line that has been drawn. One cannot make the assertion that with any degree of generality that "the angular velocity about this instantaneous center will be this amount of rotation divided by the interval of time dt, that is, it will be v/r where v is the velocity of any point of the body and r is its distance from the instantaneous center for that instant."[277] (Italics removed).

It is the same with relative as with absolute motion. For instance, there are excellent reasons for thinking that the earth is moving. Yet, this does not contaminate the fact that when we measure the course of some motion on earth we measure its course from a zero which is non-moving with respect to the earth. Physicists have recently gotten within a nano-degree or so of 0° Kelvin. Many of them believe that Absolute Zero is the point at which motion will stop. If this were so, this would be an incomplete account of its total motion. But, since their apparatus moves with the earth, in order to stop the atoms and molecules from moving without qualification, they would have to counter all the motion of the earth as such, some of which may never be known. If that were ever obtained, then total motion could be detected by an experimental apparatus. In the meantime, they will have to do with logic.

In short, there is a minimum speed (and by implication) velocity; and that is because it concerns infinitesimals only.

Motion implies something that moves, whether a mover or a moved. I know that there are people who say that this is due only to the fact that we have not experienced a motion without something moving.

The fact that there has never been a motion without something moved or a mover is not simply a case of the unfinishable type of induction. The very abstraction of motion is drawn from the experience of the initiator and initiated. Without finding a contrary instance, they are not entitled to so much as express their presumption as a possibility. The people who doubt this should consider where they are coming from.

There are times when men are unable to discern what is moving. This is when the motion within motion is too complicated, or when the usual signs of relative size are reversed. But, to repeat, of this, they can be sure—something is moving.

Much can be said about this vast, vast subject. For the present, the reader should see how easily he or she can solve the famous paradoxes of Zeno, which have troubled men for more than two millennia.

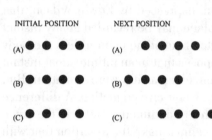

Consider Zeno's paradox, the stadium, the purpose of which is to prove that half the time may be the same as the full time. Consider three rows of objects lined up in the INITIAL POSITION. In the NEXT POSITION, the objects in Row (A) are at rest, while the (B) and (C) are moving at equal speed, but in opposite directions. After the (B) objects have each moved one space to the left and the (C) objects have, each of them, moved one space to the right, (B) will have moved passed twice as many bodies in (C) as in (A). But the time which (B) and (C) take to reach the position of (A) is the same. Therefore, twice the time is equal to half the time.[278]

The fallacy is based upon a failure to understand direction. (A) is not moving. It therefore can be set at zero. Let (B)'s direction of motion be negative and (C)'s, positive. In unit time t, (B) moved at a speed of $-1/t$ and (C) has moved at a speed of $+1/t$. The distance between (B) and (C) would be two. The absolute magnitude of each was the same; their directions were opposite. The time for one was the same as the time for the other. If people are still puzzled by this paradox, it could be because the real system is biased toward the positive, according to which the negative consists of lesser cases of its opposite.

Consider Zeno's paradox," the dichotomy," which holds that before a man can reach the end of his course, he must first cross half of it; and before doing that, he must first cross a quarter of it, and before that an eighth. Therefore, it is concluded that he can never begin.[279]

Let us see what is right about this paradox. Ignore the fact that the points on a floor are not endless. If a man attempted to cross a floor by covering first 1/2, then 1/4, next 1/8, after that 1/16, he could never complete it. The series makes it impossible to complete the goal, and it would be fudging to claim that since he almost reached it, he did.

This is sound as far as it goes. If one did attempt such a thing, one would never legitimately cross the space. What Zeno did was turn this series around and begin with the man standing on the limit outside of it.

The truth is that, even under this assumption, the man could cross the floor. The first reason is that motion exists. That is known to everyone who can move his or her fingers across the page and enter the information into his mind, having first taken the time to acquire a certain skill at reading English. (This point will be elaborated in the discussion of the next paradox.)

Second, let us consider the reasoning which seems to place a difficulty in the path of intellectual advance. Let us begin with the man who had managed to cross so much as half the floor and show why he can complete the rest of it—after which Zeno's reverse series will be considered. The size of the steps he takes need not be (and usually are not) in a geometrical ratio to those which he had already taken. Subsequent steps are usually determined by how quickly he wants to cross the floor. What remains after each step is not the fractional part of the whole already crossed, but the remaining difference. And once more, this shows the distinction between division and subtraction.

The process can also be described by division, but it is one in which the denominator changes with each step. Suppose the floor were twenty feet long and the man's pace was two feet at a time; after completing five steps, he will have reached the half way point, but his next step at that pace would complete 1/5 of that remaining, leaving a distance of eight feet to be crossed. The even step after that would be 1/4 of that remaining; the one after that, 1/3 of the six remaining; the one after that, 1/2 of the four feet left; and then the completing step, 1/1, leaving nothing more. What has been performed is subtraction, rather than division, leaving 0 distance to cross. The full series, starting from the beginning, rather than the second half is different, but it too illustrates the same point: 1/10, 1/9, 1/8, 1/6, 1/5, 1/4/, 1/3, 1/2, 1/1.

Now, to consider why the man can begin when he stands in the place of the limit never reached by . . . 1/32, 1/16, 1/8, 1/4, 1/2, 1/1. If he tried to make his pace fit that series, of course he could not so much as take his first step; motion would be impossible under those circumstances. But he does not have to do this. He can simply step over them.

This shows that so far from interpreting motion, the limit prevents it. The geometrical tangent measures the curve by resting outside of it, but the limit is not only outside the motion, but prevents it from happening at all.

If one drops the initial assumption and recalls that the points in the path are not unlimited, then, providing that the man took the first step, he could complete it anyway.

This also, to borrow a phrase from Shakespeare, "gives the quietus" to the idea of microscopic infinity: it could never get started.

Another of Zeno's paradoxes is "the arrow." An arrow in motion is either at rest or not at rest. If the instant is indivisible, the arrow is incapable of motion; if it did move, the instant would be divided, which is contrary to hypothesis.

The answer is that motion cannot be reduced to non-motion. At that very instant, there is a change in position. It is not the infinitesimal of time that is divided; neither is the motion divided. Instead, some locations are crossed;

motion takes place from at least one spacial infinitesimal to another one. Although the space is divided, the motion remains undivided. The faster the speed, the more infinitesimals are instantly passed. The time at which it takes place is infinitesimal, but at that instant, infinitesimals of space are traversed. In instantaneous speed, there is a beginning and an end, but no middle. After immeasurable quantities of successive instants, a finite distance is crossed; the greater the motion, the greater that spacial interval.

The reason for the difficulty in understanding motion as something which cannot be reduced to divisions in space or in time is the confusion of the means of measurement with the thing being measured.

The visual system does not register a pure continuity. When we watch the flight of a bird, our retinas do not mark each and every point of its flight path. Neurons are fired at specific points along the trajectory. If our visual perception were broken into its smallest parts, the trace of the bird's motion would be a like a strip of film with its many tiny stills. We experience motion, because the same bird is at various places at different times. We do not reason that motion must have taken place because of the law of contradiction. To the contrary, we *induce* this law from the particulars of our experience of motion. We are able to formulate that two or more events cannot occur at the same place and at the same time because we know that space, time and motion are not the same, but different. To seriously attempt to think otherwise would be like trying to remember things wrongly. It might crack our capacity to apprehend.

Our consciousness of motion did not derive from a comparison of percepts. Like that of time through memory, it is innate. Our mind is designed to interpret the perceived change in the placement of the same of object as *motion*. To doubt its reality is foolish. The bird had to move in order to be at different places anyway. Even the wildest skeptic would have to admit that something had to move.

Space, time, and motion exist independently of one another. The fact that one cannot measure time without referring to observations of some matter in space does not make the former dependent on the latter. Even if the instruments were perfectly reliable, the instant would be too short. Neither would matter affect space, for the latter is empty and does not move aside in order to accommodate it. Various authors have shown that Albert Einstein could not consistently argue against the existence of such a space.[280] On the same page in which he wrote that "there is no such thing as an empty space, i.e., a space without a field", he also stated in a footnote that "if we consider that which fills space (i.e., the field) to be removed, there still remains the metric space."[281] Void the "field" and what remains?

The last of Zeno's paradoxes is the Achilles. The argument is that Achilles can never overtake a crawling tortoise which started ahead of him, because whenever he reaches the point where the tortoise was last, the latter has moved on. This one, the reader can solve.

In short, motion is a fundamental fact of existence. Space and time are the background for the existence and motion of material objects. If this were problematic in the way so beloved by 19th and 20th century skeptics, then so would be the empirical studies of which they are so fond—also the mathematics and logic they toasted.

CHAPTER XXV

CONCLUSION AND SUMMARY

In this book, the infinitesimal is not used as merely a convenient tool of analysis which can be dropped after it has accomplished the task which the mathematician has set for it to do. Rather, it has been shown to exist, independent of human power of conception. The number line includes both the continuous and the discrete without slighting either.

Quantity is a fundamental notion which is understood in opposition to quality. Attempts to produce concepts of quantity completely devoid of any connection to quality are doomed to failure. The two are fundamental correlatives which require each other.

The discrete and the continuous are quantities. The first is countable; the second is measurable. Neither can be reduced to the other on the finite level. Whether the first mathematician began with counting or measurement, the other kind would have been discovered sooner or later.

Set theory over-symbolizes. Under that theory, continuity is reduced to the discrete by treating number classifications as if they were turnstiles, clicking in their members as the latter enter the box connected to it; the various kinds of numbers are conceived as being, each of them, set in their own boxes, enumerated up to infinity, although no one doubts that this is impossible. That contradictions and amphiboles should develop from that is not surprising.

Quantity and number are not the same. A number is a symbolical representation of a discrete quantity which has been identified, according to some system of enumeration. The decimal system is based upon the number ten; it is not the only one possible. An earlier system was based upon the number twelve. Although the decimal mode of calculation has a natural analog in the ten digits of the hand, that other system has more factors, six instead of four.

Since every number in every system is finite, no N, however large, can be infinite. Numbers must be distinguished from quantity. An infinite number, we cannot speak of; an infinite quantity, we can.

Everybody comes to the idea that there must be something beyond counting. Some reject it altogether, because they say that everything is just so much. Others do not deny it, but take the position that they will only consider the finite, since they know that they are equipped to handle that. Still others realize that it exists and that although men may not be able to handle it directly, they must, at least, be cognizant of it if they are able to make full sense of the finite.

The numerical system is incapable of calculating an actual infinity, since every element must be finite and no accumulation of them can pass out of that metaphysical state. Set theory confuses quantity with number. It treats what is actually a human contrivance as if it were something finished and present from eternity.

The number line, by contrast, contains both the continuous and the discrete without attempting to reduce one to the other and without pretending for an instant that it is actually infinite. It is found in the points along an already identified length. A specific number—+5, for instance—is not some point, but the distance from that point to the beginning of the number line.

A common material ruler has this characteristic. Even though the edge may be uneven, there are within it from zero, or the beginning, to the inch mark, an immeasurable quantity of points.

Knowledge of this fact points to the existence of the infinitesimal. It is nothing more or less than a location in space. It is the smallest existent; nothing can be smaller. It has no shape. If they were round, there would be places where they were no points; if they were cubical, they would occupy more than a single point; each corner would located in separation from each other. The same with any other shape, for shape is a finite characteristic.

Although this reality cannot ever be sensed, the mind can become aware of the fact that it must exist. In chapter VII, it was shown that it cannot be denied without obliterating part of reason. To talk about it as being useful as a ladder that can be thrown away after that level has been reached is intellectually irresponsible. As was shown in our study of the calculus, its denial produces mistakes.

The number line is anchored in the reality disclosed by the senses. It is made possible by the linear properties of space—themselves slices from the three-dimensional world. Any real, veritable, or absolute magnitude in the number line can be related to the unit, either exactly or approximately. The emphasis is on *can be*. Prior to the identification of the point as part of the number line, it has no relation to any such human contrivance as a number. Contrariwise, no recitation of digits is a number, until it has been related to that line. The exact point does not have to be physically known, not even a region needs to be indicated, dextrously. But a rule must be connected to it through the completion of an actual calculation.

Direction is a fundamental attribute of a line. It is, therefore, not surprising that this important fact should be associated with numbers. There are three ways to do this, reflectively through real numbers, non-reflectively through veritable numbers, and through their combination, the imaginary lines. Exploration of the properties of veritable number will further the advance the science of mathematics. Its use in physical science is like a path yet to be hacked through a great forest.

It is vain to suppose that numbers can be completely independent of geometry; geometry can never be simply a part of physics, as the Objectivists would place it. In the case of the transition from the theory of limits to set theory, what began as an attempt to stay within finite human terms has turned into an escape from reality.

Suppose that it were possible to "construct" a number system that was continuous and that it was a free creation, made without the number line. The general concept, continuous number system, would be a genus with two species, the first being the number line with points next to each other and the second, a number system with no beginning (unless zero was put in—in which case, it would not have an immediate successor). The two systems would stand forth as alternatives, but would they be equal?

The number line championed here has some possible locations—the positional points—that the other one would not have. The free creation cannot contain the positional points, for men are incapable of numbering them, even within a small interval, such as that above or below 0. Contrasted with the original number line, the free creations posited by Dedekind, *et al.*, would be discontinuous.

There must be a next-to. Otherwise there is no real continuity. It is not a matter of what we human beings wish. If there were no next-to, then to move from one point to a second point near-by, we would have to jump cross a gap. What lies within the gap?

If there is a third point on the other side of the second point but next to it, the latter becomes a middle point and the third one, the endpoint. But if there were gaps between the points, there would in effect be edges everywhere—in which case, it would not be continuous. For strict continuity means to be without gaps within it.

The answer from the advocate of 20th century mathematics is that given any interval, another one smaller could be found. But what about a point just in excess of the former number? They would answer that there could not be such a thing as a "just in excess." That between that one and the first number, they could come up with another. But could they? If the number line is continuous, they could not, for continuous means without a break. But then, their contraption would not be continuous in the original sense of the word, because the new number would have to be invented when needed. If it were there already, then they would be back to the number line with its already existing continuity made up of points, any one of which could be identified as the terminals of a number. But if these points did not already exist, then their act of division would really be one of creation in which the components of the interval to be divided did not exist until after the act of division had taken place. And if that is the case, then what would the interval be prior to division? Whatever it was, quite simply, it could not be an interval, for that would imply that there were something to be divided, that parts, aliquot or irrational, already existed. It would be as if the very act of division established not only the denominator, but the very possibility which was to be divided. But that would not be a division.

Even more crucial, there would be nothing analogous to that aspect of subtraction which is not contained in division. This book has shown otherwise, i.e., that such subtraction must exist. That being the case, the pretended infinity based upon continued division does not stand. Considered as an operation in and of itself without reference to infinitesimals, division is inherently incompletable. Subtraction is completion. If there were no completion, there would be no incompletion.

The existence of the spacial infinitesimal is inescapable. It is not given us, innately, like time through memory. Nor is it a direct implication of organic experience like our knowledge of the existence of space. Rather, it comes to us as a further implication of reasoning about the continuous and the discrete, two fundamental abstractions easily traceable to the senses.

Both set theory and non-Euclidean geometry attempt to abstract away from the senses, the former for the sake of a strange sort of perfection, the latter in order to make the contradictory seem compatible. The answer, however, is to abstract by means of that which has been perceived, to analyze but not lose consciousness of its content.

To accept infinitesimals is not to deny sense. Their acceptance is an affirmation of the senses, for it is through reasoning about them that one comes to the realization that infinitesimals must exist. They are not to be dismissed as mere projections of concepts upon the external world. On the contrary, it is only after the world that is disclosed by the senses is affirmed that the necessity of admitting the existence of something which is not subject to division is understood. With the infinitesimal in mind, knowledge of an immeasurable infinite comes into focus—that and more. Far from standing in the way of obtaining what is beyond them, the senses provide a way.

In truth, the deniers of sense experience are obliged to admit it by the back door. The objects of the world, the things we encounter, cannot be replaced by pointer readings or digits on a screen. To receive them, we must trust our senses. Beyond that, if all that was required were the readings, then we could simply set them arbitrarily ourselves and skip the ultimate recourse to objectivity. After that, the statement of Macbeth in the great play that the world is a "tale told by an idiot, full of sound and fury, signifying nothing" would be too kind; there, reference is at least made to outraged nature.

With the recognition of the infinitesimal, the ancient puzzle about the nature of the non-rational in mathematics is solved. An irrational magnitude is not a breach in continuity, as Dedekind supposed.[282] It is just that the possible magnitudes within a unit cannot all be understood at that level. The unit is finite, but its ultimate components, the infinitesimals, are not. Stated more technically: *It is because the difference between the length of an irrational number and that of the rational magnitude nearest to it is below the level of the finite that the former cannot be identified as an aliquot part of the unit.*

At the level of the infinitesimal, there is no attribute, no signature, that would distinguish a position that is the upper point of a rational number from an irrational number. Infinitesimals differ only with respect to their position in space; in fact, that is what they are—positions. If the unit were of a different length, or began at a different point, the same infinitesimal which was the end point of an irrational number might be the endpoint of a rational number.

Finally, there are the positional points, which are so near the beginning of the unit that they cannot be the upper point of a magnitude.

Mathematicians who fail to understand infinitesimals often come to contradictions. An example is the attempt to show a 1-1 correspondence of the points on a line segment with an area formed by the rectangular square of that line segment; the points cannot possibly match. Earlier mathematicians, like Girolamo Cardano in the 16th century, understood that it was fallacious to compare areas with lengths.[283]

There are two kinds of infinity, potential and actual. Knowledge of the first originally took rise from the awareness that no matter how many things one could notice, there could be more. Knowledge of actual infinity came about through the contemplation of continuity.

We have found that modern mathematics has dropped an ancient distinction between the two kinds of infinity, attempting to reduce it all to actual infinity, the very kind which is conceptually most difficult to handle. In so doing, they have made a hard subject more obscure than it needs to be.

Just as the idea of nothing but finitude quickly loses creditability, so it is with potential infinity in the absence of actual infinity; ditto the notion that they can be conflated. The idea of escape through the limit also meets defeat. The infinitesimal can be used consistently throughout mathematics, and it leads to new knowledge.

Set theory is mistaken. It is not the infinities represented potentially by numbers that is important, with natural numbers representing one type of infinity, rational numbers another type, and irrational numbers a third. What is important is its type, i.e., whether it is real or veritable, rational or non-rational. All are beyond numeration, but defined with reference to the unit. Sometimes, one instance suffices to conceptually identify a type. In the absence of a particular length being designated a unit, the infinitesimals along a line are all the same, distinguished only by their space, i.e., by themselves.

It is by reference to the finite that men order or rank the infinities. Contrary to Cantor, they do not know how to rank them in terms of what they are in and of themselves. The reason for this is that men come to know that there must be an infinite only by the problems raised in handling the finite. It is this double redaction—the referring back from the infinite to the finite and the finite to the infinite—that brings about depth of understanding. Mathematics, like everything else men do, is tied to the human condition.

Actual infinity may be divided into two types: the immeasurable and the endless. Consider the first kind. The immeasurable exists within the unit. If one took the points at the base of a standard square, their quantity would be "$\mathfrak{U}$." If one took a quantity equal to these points and then made each of them one unit in length, the line would be "$\mathfrak{I}_{\mathfrak{u}}$." But these would all be limited.

The old idea of inexhaustible numbers within the unit has been shown to be false. In its place is a unit consisting, ultimately, of infinitesimals, all of which are positions and are, therefore, located in reality, not just in the imagination. This truth stands out, despite the fact that the infinitesimal is humanly immeasurable, and for that reason beyond countability. Only the units and their identified parts can be counted.

Consider now, the second kind of actual infinity, the one without limit. Its symbol is "$\mathfrak{E}$," for endlessness. Space is the most obvious example of this second type of infinity.

The discoveries of the infinitesimal and the infinite are a type of thought that is higher than the induction discussed in Chapters VI and XII, wherein an abstraction is drawn from the existents found in art and nature and then solidified into an over-arching concept. It is also way beyond the kind of empirical induction which is hardly ever complete; it is not by counting concrete instances that we call what we can directly perceive and reason about the "finite." Its existence is necessitated by the careful consideration of what is implied by the world disclosed through perception. Although both deduction and induction are involved, intellectual discovery is crucial. This last is not a breakthrough, but a confirmation of the known world, a recognition of consistency which calms the waters of thought.

It is not a mere "coincidence" that numbers apply to material objects. Once the infinitesimal is understood, the ideational in geometry is no longer in conflict with the physical, thereby cutting off the excuse for a resort to the pragmatic. Approximates are seen as exactly that, and not intellectual incongruities to be explained away through such devices as "cuts" and the ellipsis.

Measurement is based upon lines; these can be located in space; they can take it up. Counting is based upon inspection of the objects of the world. Length, width, and depth are fundamental abstractions from the facts of existence. The idea of duration is an application of lines to the passing of time. Motion has its path across space, which also is discernable in numbers.

The burden of proof remains on those who proclaim the existence of an additional spacial dimension. Arguments based on consistency are unwarranted, for, as everyone knows, a person can be consistently wrong. Merely because the denial that parallel-lines-cannot-meet may be made without destroying the entirety of geometry does not mean that there is room for non-Euclidean geometry. Their argument is a complete *non sequitur*. Within proper geometry itself, non-intersecting lines are not always parallel.

What keeps the folly from being recognized for what it is? It is the hidden premise, widely shared by so many, that all is one, that everything is wrapped up in everything else. Riemann may have concurred, for he once wrote that "a complete, well-rounded mathematical theory can be established, which progresses from the elementary laws for individual points to the processes given to us in the plenum ('continuously filled space') of reality, without distinction between gravitation, electricity, magnetism, or thermostatics."[284]

Third world paganism is replete with it. So is Marxism. But one can also find it in ante-bellum America. Ralph Waldo Emerson, the early 19th century essayist, included these lines in one of his poems:

> Line in nature is not found;
> Unit and universe is round.[285]

Concealed within the set theory's double focus is the suppressed premise of omniscience. The set does not merely refer to that to which it corresponds, as with an ordinary concept, but the referents are in some way contained within it. When Cantor supposed that his infinite numbers all existed in the thought of what he believed to be God, the presumption of complete knowledge at least made some sense. But when employed by this-worldly theorists, the premise still remains, however thickly disguised by overwhelming symbolism. One suspects that what such theorists hypothesize is not some personal Deity that is omniscient, but a mental force inhabiting everything—in short, pantheism. Once more, in their theories, all is one.

Recognition of the existence of the infinitesimal does not result from playing with alternative axioms which are supposed to mean something simply because they can be stated without contradiction; this fundamental reality is recognized from the limitations of sense perception. The questions which bring about its recognition are drawn directly from experience, not from vain imagination.

In accordance with this realization, further questions come up. One is this: What is the division between the finite and the realm of the infinitesimal? Is the answer that it is simply that which lies beyond the reach and range of the senses and the machinery used to extend it? Such an answer is only partially correct. It is true that the practical finite is the present degree of knowledge. *As long as we cannot penetrate beneath the level of irrational magnitudes, we know we are still in the realm of the finite.* But we do not have to know the final limit of that ability to possess a sufficient answer. The infinitesimal has no shape; it is neither round nor does it have corners. Its realm is below both the practical and the ultimate finite.

Another question: Did the finite issue out of, or, to use the popular word, "evolve" from the infinitesimal? Since space is totally homogeneous, there is no difference from which the differentiated could emerge. The first line could not have been laid down as one point connected to another; infinitesimals are not hooked. There could not have been an evolution from the lower. It had to exist from its beginning as finite, not some inchoate missing link.

The two are radically different from each other. Infinitesimals cannot be measured. The unit *can*. At the level of the infinitesimal alone, there are only multiples of the indivisible and rational subdivisions of this multiplicity. At the level of the finite, once the upper and lower bounds of the unit have established the interval between them, rational fractions do not suffice; a rational subdivision is fully intelligible as a part of the unit, i.e., it can be stated as a ratio. But there are determinations which are not fully explicable in terms of these parts. The pagan doctrine that all is one is false from the start. Unit and universe are not round.

The size of the unit is determined by its smallest rational number, itself the reciprocal of the largest whole number. The length of the number line, itself a potential infinity, is dependent, not just upon the upward advance of its successive heights, but upon the growing size of its unit. In short, this line snaps forward.

Then there is the issue of the sturdiness of the unit. The linear unit may be subdivided until there is nothing left; this is because the quantity of infinitesimals within a unit is immeasurable. But there is a size beneath which the circle cannot attain without disintegrating into mere infinitesimals.

Ultimately, the absolute effective minimum length would depend upon the smallest division of energy. The two types known to modern science are matter and light.

But this would not be the theoretically minimum unit. To be that, the length must be great enough to handle as a rational subdivision the smallest possible interval. Just constructing and putting into operation a computer which can type out zeros as long as it can run does not "count." It does not express any reality.

Because of what we recognize as the inherent limitations of the finite, we are certain that the infinitesimal exists. We can even learn some facts about it. For instance, while the infinitesimals of space do not move, they are everywhere at once. But the finite can move. They are not everywhere at once. These and other facts and contrarieties we can learn.

Because of this recognition, we can solve the ancient question of the many and the one. The individual locations, the infinitesimals of space, are the ones of identity, while the spacial units are the ones of unity, capable of division and distinction within.

Because of our recognition of the existence of the infinitesimal, we are able to distinguish between the two kinds of actual infinity; with it, we can conduct *division* by zero.

Because of this recognition, we are also able to determine the minimum angle—an opening of one infinitesimal. Any angle greater than this can be split. If there were an angle of two infinitesimals, the maximum possible quantity of lines would be half as great; an angle of three infinitesimals would contain one third as many as the least.

The number of infinitesimals in the minimum circle would need a radius long enough to support π.

Knowledge of the infinitesimal enables us to understand the indispensable role of the tangent in the derivative, that it matters whether one line is on another line or through it. Forgetting this has led men to think that they can throw away the geometric ladder whereby they have ascended and stay aloft on a cloud of algebraic manipulation. Similar mistakes led to the inherently inadequate limit approach in the calculus and the faulty analysis of the integral into rectangles.

Knowledge of the infinitesimal reveals $\triangle y$ and $\triangle x$ in the derivative to be exact amounts, whether finite or sub-finite—not enigmas hovering imprecisely over nothing.

Knowledge of the infinitesimal allows us to understand how a number line simultaneously calibrated by two incompatible systems can be combined into a third without contradiction.

Like the infinitesimal, the infinite is beyond human ken. And it is also more difficult to imagine, since it is not indivisible. Yet, we cannot deny its existence. But we can work with it. We know that it is not all the same. We can distinguish anything we define as $\mathfrak{I}$ and $\mathfrak{E}$.

Time is not an extralogical element. The supposition that it is produces an ideational system which is ultimately fallacious, as was seen with the way continuity was handled in the two previous centuries.

Unlike space, there is no realm of existence in time apart from the instant—no finite in contradistinction to the infinitesimal. Once an instant is gone, it does not go to someplace else; to the contrary, it then ceases to exist. If one were to bring up the same physical objects and processes, even the identical psychological reactions of an observer to an earlier event, it would make no difference—the uniqueness of the moment would be gone. This fact transcends such mundane questions as to whether or not some laboratory would be able to discern any difference.

Neither does the world float in time. Time has no arrow, either; it does not move forward. Each instant is replaced by its successor. There is no breach as one replaces another. Infinities of the discrete through replacement—that is what it is. Yet, it is also continuous; it is without gaps. But there is no contradiction in this, since it is discrete and continuous in different respects. This thought is astonishing!

The only mathematics which can represent the instant are the natural numbers, or better, numbers of identity. One can count; that is all. Fractionalization and the rest are impossible there.

That the passage of time, duration, is represented in spacial terms is of immense theoretical importance. The minute, for instance, is set by some material motion experienced or measured as regular. Day and night are astronomical phenomena. In a careless mood, a person might think we can bundle the past in a manner analogous to the way that the finite encapsules the spacial infinitesimals. The difference, however, is crucial: whereas the distinction between the finite and the infinite is fundamental to space itself, the difference between the instant and the minute or between the second and the instant is not rooted in the nature of time, for neither past nor future exists. Yet, a line and the spacial infinitesimals in it both exist simultaneously.

The infinitesimal of time is not accurately represented by anything spatial. Even the point is only an allegory; the instant, since it is the same everywhere, is not particular like the point.

Space and time are fundamentally different. Unlike space, we do not become aware of time through perception of the outer world, but by memory.

Even $\mathfrak{C}_t$, or eternity, is not time, but symbolizes something that never ceases at any instant, i.e., always persists. It is true that the fact of time does not die, even though its instants do, but it has no being apart from the instants themselves; it is simply the fact that *is* really is. But that fact is not the only kind of eternity. God is eternal.

Commonly, modern thinkers try to pair the dimensions of space with duration to form what is called a "4-D" system. This appears to be plausible, because so few understand either space or time. Einstein with Minkowski attempted to do this by writing an expression in which the first three terms stood for the dimensions of space and the fourth term had $\sqrt{-1}$ for a factor.[286] But as we have seen, there is nothing mysterious or ineffable about this number; it is an element in the veritable number system, something was not discovered until almost the end of the 20th century. This number has nothing to do with duration.

If one were to try to pair space with something better understood, this idea would hardly be given any importance whatsoever. Suppose one were to pair x, y, and z with p, where "p" equals the morning price of Kansas wheat over 20 years; and would to try to sell others on the idea that this was a four-dimensional existence. Imagine that one were to contrast with this "p" a three-dimensional spiral. Then any particular juncture of x, y, z could be reduced to a point which then could be "mapped" in another coordinate system in which each of these points could be compared on one axis with some particular daily wheat price. Doubtless, some consistent equations could be made, relating the spiral to the wheat price cycles. The only trouble is that far from disclosing a portrait of the fourth dimension, it would only relate what is not causally connected to some figure. The resulting drawing would be more like a modern expressionist painting than it would a scientific exhibition.

Despite its uniqueness, the same time exists everywhere. Nothing can start anywhere in space while part of an instant somewhere else is unexpired. This is because the instant is indivisible; there is no place for one instant in one galaxy to overlap the instant of another galaxy. That and not some agreement between clocks everywhere guarantees the instant's omnipresence. That is why time is both absolute and universal.

Neither material, nor mental, yet time exists. Truly, it is something to wonder about. Twentieth century scientism with its cheap division of existence between the psychological and the physical could not find it.

Now, let us consider motion. Just as there is an atom of space and an atom of time, so there is an atom of motion. ("Atom" is used in the original sense of being uncuttable.) The atom of space is reachable through subtraction, at least theoretically. The atom of time is unreachable even by subtraction and is known only by reflection upon an idea disclosed through the innate process of memory. The atom of motion is known through an implication arising from the nature of the means by which it is measured, i.e., space and time.

Although motion must exist across a finite number of infinitesimals in order to get anywhere, its hallmark is transportation across the finite. It is motion that gives us knowledge of the existence of intervals. This does not mean that intervals could not have existed in the absence of our motion. Of course, they could. But we are born in media res, in the midst of things.

The construction of a number line is an action. Zeno's arguments against motion are like the case against the infinitesimal. With the former, it is supposed that the first step is prevented by an asymptotic process; with the latter, it is supposed that there can be no final point, that everything is in an infinite hole.

In Chapter VII, two squares were presented, one of them filling the interior through continuous addition of horizontal lines, the other never filling them. Without finite motion, neither square could have been completed. The dividable square, like the asymptote, is a potential infinity. It could be completed only after multitudinous divisions beyond human ability.

When the veritable and the real are brought together in the imaginary axis, motion takes place. Motion is also revealed by the interchange of the real and the veritable as the radius vector sweeps around the unit circle—and beyond.

The -1 inside the √sign is veritable, not real, because the imaginary axis is as intrinsic to the geometry as is the real axis.[287]

Knowledge of the existence of motion allows man to mentally connect the finite and the infinitesimal. Behind any finite speed must lie the fact that so many spacial infinitesimals are crossed in an instant. Time itself cannot be finite, for only the present exists.

The frequently quoted statement of Aristotle's that "we measure the movement by the time, but also the time by the movement, because they define each other"[288] is not true. (1) No such measurement would be possible without consideration of space. (2) Time and motion do not define each other. Far from it. The consciousness of time originates in memory alone, while the consciousness of motion requires the active use of other faculties besides the ability to remember.

The primary kind of motion is speed. And speed is not a ratio between distance traversed and a time interval (duration). Rather, they are the means by which this fact that exists independent of them is measured. Speed cannot be the measure of itself. Furthermore, duration does not exist as such; it must not be forgotten that the past only was. Duration is an analogical projection of the idea of space upon the idea of time. The finite measurement of speed, therefore, uses the attributes of space, not only in the numerator but also in the denominator; this is because the reference motion used to measure duration has spacial aspects.

It follows that space alone is the measure of itself, which proves again that it must exist, though it cannot be perceived.

Space was not laid down. "Laid down" is a spacial phrase. Space always was. An infinity of infinities of infinities, it is simultaneously present always. This shows the difference between space and matter. A material object is not a bundle of lines. Large or small, it occupies space immediately by its very presence. Since space is empty, nothing is pushed aside for it to occupy some place; and even if there were such an aether, it, too, would occupy space. This was another mistake that Descartes made. He thought that matter was just extension, i.e., that space and matter were really the same thing, rather than different facts.[289] This mistake was also carried onward by Leibniz.

The attempt to rely on analysis, ignoring origins, led to the idea that it didn't matter whether the tangent lay on the line, as the classics taught, or was in it. They figured that a symbolic analysis divorced from sense would take care of everything. Yet, with the extra infinitesimal, the tangent usually indicates

222   PETER F. ERICKSON

something static; without it, the neo-tangent can assist in calculating motion. This earth-shaking innovation of Descartes has gone all but unnoticed.

This is not to disparage those who love analysis. Experience has shown that analysis can handle many complicated problems more easily than can geometry alone. But the latter is much more suited to the consideration of the fundamental. And it is about fundamentals that this book was conceived, written, and published.

Now, let us say a few words about relative versus absolute motion. This was the great intellectual issue of the century just passed. Instead of burying ourselves in the welter of meanings given to the key words, let us define relative as Einstein did, to wit, that there is "no such thing as an independently existing trajectory, but only a trajectory relative to a particular body of reference."[290]

This means that *absolute motion is that which has a trajectory independent of any observer.*

Suppose that a battery powered traction vehicle with magnetic wheels were programmed to move across an iron disk a mile in diameter from a spot marked on one side of a perimeter to a spot marked that is diametrically opposite. Let it begin to move. During any particular instant, it is covering one or more infinitesimals. This is a fact, regardless of whether or not its velocity is constant.

Suppose that the disk were rotating. So long as the disk was not turning so fast as to throw the vehicle off the track or overturn it, it would still be moving from a mark on one side of the perimeter to its opposite on the other side. This would be its course on the disk.

Suppose that there were an observer K on a platform fifty feet above the disk. Providing that the disk were not rotating too fast, the observer should be able to see the heading of the vehicle. Despite the fact that the disk itself is rotating, the movement on top of the disk is still from one side to another. To what extent would its ability to reach its destination be affected by the fact that what it was on was rotating? There would be an uneven wearing on the wheels and tractor parts; there would be some breaking against a slide. But if it was set to reach the other side, its mechanism could adjust enough to hold course.

Suppose next that both K and the rotating disk were tumbling together. Let K be strapped in so that he could continue to observe the vehicle moving across the disk. Since the vehicle's wheels are magnetic, it might still be able to continue its heading, albeit by applying more force against the torque. Suppose it does, and that K is still able to observe its motion along the surface of the disk.

Suppose next that there were another observer. Let us call him K'. Suppose K' could also see where the vehicle was going. The Einsteinian would have to say that since the two observers agree, this is what is happening.

Now suppose that both the tumbling vehicle-on-the-disk—K system and K' were rotating with respect to another point. Suppose further that standing on a platform at the center of their joint motion was another observer, K″, but that the latter was so far away that he could not discern from the compounded motions what the robot vehicle was doing. The Einsteinian would have to conclude that the vehicle was moving from one side of the perimeter to the other for K and K', but not for K".

But such a conclusion would be contrary to hypothesis, which is that K" was too far away to see clearly what was going on at the surface of the disk. If the Einsteinian were consistent, he or she would have to argue that there is no such thing a bad observation site. (Backers of scientific institutions could keep this is mind the next time they receive a request for additional funds for a "better" telescope.)

If a boy with binoculars is unable to see a star that can be discerned at Mount Palomar, it can only mean that they cannot agree. If there is no trajectory independent of a framework of motion, the question of the star being where it is regardless of what both observers discern is not to be raised.

To this, it might be answered: If the boy should take a trip to that observatory and look through its scope, then he could agree that it existed in that context, but when he came home, that it ceased to exist.

The rejoinder: If the boy's eyes were too weak to see the star even with the assistance of Mount Palomar's mighty machinery, then the star did not exist even when he looked through the sight. By this sort of reasoning, after all the objections and counter objections had been gone through, the strict relativist would have to conclude that there is no such thing as an inept observer. To get around the conclusion that they must take into account an underlying reality transcending perspectives, they would have to resort to a kind of majority-rule-intersubjectivity.

Relativity is wrong. The boy is incapable, and the vehicle is heading toward its predetermined objective regardless of whether or not anyone is watching. Its motion on the surface is absolute. Any rotations or revolutions that the disk is simultaneously undergoing which might impede the course of the vehicle will have to be compensated for by the servo-mechanism. Any too faint to change anything will require no adjustment.

But this is only a backwards statement of the first law of motion. An object continues on its course, unless disturbed. At any moment, its step is the net of uncounted, perhaps countless causes. Every motion described is absolute; it is what it is, independent of whether it is observed or not.

Some might argue that there is an important exception to this, i.e., when the mechanism used to perceive it interferes with the vehicle's motion. But such a case would only change the net. The servo-mechanism would react, and the altered motion which resulted would be as much an absolute as what would have taken place in the absence of the interfering observation.

Henry H. Lindner, in an intelligent article in *Physics Essays*, argues that Einstein's subjectivistic theories served to give the science of physics a way of overcoming discrepancies during a time when it was in theoretical turmoil. As a young man, Einstein had thought that this was enough; in fact, he had conceived of science itself as mere intersubjectivity—what he called the "comprehension, as complete as possible, of the connections between sense experiences . . . ."[291]

Later in life, however, Einstein realized that something more substantial was needed and tried to provide it. In the beginning of his career, for example, he followed Ernest Mach who taught that atoms were just "convenient fictions." But later on, he was to write that Mach "condemned theory on precisely those points where its constructive-speculative character unconcealably comes to light . . . ."[292]

Einstein went further. He finally admitted that science has to be based upon a belief in an "external world independent of the perceiving subject." [293]

Dr. Lindner objects that Einstein was still an advocate of consciousness rather than existence. Lindner wants a physics which moves in the direction of what he calls "Cosmic theory: the attempt to explain our existence and experiences as caused by observed and unobserved entities and process." [294] He holds that Einstein was still wrapped up in a thought which was other than about reality.

"Physics," Einstein wrote, *"is an attempt conceptually to grasp reality as it is thought, independently of its being observed. In this sense, one speaks of 'physical reality'."*[295] (Italics added.)

To this, Lindner objects with great eloquence: "Einstein believed in atoms and his other constructs, but only because these ideas worked to make our experiences intelligible. Einstein reached out to a reality beyond our sensations, but that reality consisted only of more ideas. Beyond reality as it is observed he believed in reality 'as it is thought.' He did not posit a physical Cosmos, nor did he produce any physical hypotheses."[296]

Lindner says that instead of having motion relative to some observer, regardless of how trivial, it should be relative to large bodies, even to the movement of the Cosmos itself.[297] But this is still a kind of relativity, this time a materialistic variety.

In a certain regard, Einstein with his goal of a physics that can grasp reality as it is thought came closer to the truth. The world as disclosed by the senses announces and gives notice of its existence. We have found that realizing the truth of the infinitesimal arises from contemplation of the necessities inherent in basic perception. Once one has seen that to reject it is to reject experience itself, then one can understand that it is an important disclosure of reason itself. That thought brings us absolute space. Memory announces the existence of a time that was not. Contemplation of that gives us knowledge of the infinitesimal of time, which is the same everywhere. Although in a manner less intellectual than time, the perceptual system gives us an innate awareness of motion. Through a process of reasoning involving its logical connection to the other two, we come to the infinitesimal of motion.

Since the knowledge of time comes from an innate idea and the knowledge of motion through an inborn recognition system, and the knowledge of the external existence of space is the only reasonable conclusion to be drawn from what is experienced through the nervous system, any physical theory drawn exclusively from the results of measurement will be off-reality to the extent that interpretations are made without taking into account the metaphysical priority of that which makes any such data possible. That is why relativity is only a brilliant intersubjectivity, a strobe flash which only makes the darkness greater than before.

The trajectory of the vehicle across the diameter of the iron disk is independent of any external observers. The pushes and pulls of all the extra rotations, revolutions, and translations which the servo-mechanism of the vehicle must brake against in order to maintain its course provide the extra context for the system, regardless of whether or not there are any observers present.

If there were no compensatory machinery, whether or not the battery-driven-vehicle could maintain its programmed trajectory, or even stay on the iron disk, would depend upon what exactly those forces were. In either case, its trajectory is what it is and not something else. As for those motions that do not affect the movement of the vehicle: it does not interfere with them, nor do they interfere with it. As Morton Mott-Smith put it: "Each motion is fully carried out regardless of the existence of the others. This is the principle of the independence of motions. It is nothing other than a strict application of the law of Galileo."[298]

It is wrong to speak of a body as moving relative to space in a sense analogous to Einstein's relativity. There is no way anything can move without being in space. The dynamic term "moving" is as much a spatial relation as are static terms like "in," "on," "off," "above," "below," and the like. A distinction must be made between Relation and Relativity, the latter of which has to do with an observer's perception of it and how two or more observers can correlate their measurements. By contrast, a relation is a fact concerning two or more other facts.

The relations between the vehicle and the disk are that the former is on the latter and moving across it, and the disk has something on one of its large sides that is moving across it. Even when we take into account the jostling and alterations of the programmed course caused by the other motions mentioned above, it still has its trajectory from one side of the disk to the other. Whatever its total motion may be, the vehicle is at every instant somewhere in absolute space. An unmoving platform is unnecessary.

Knowledge of the three kinds of infinitesimals refutes the central intellectual idea of the 20th century, namely that science must always be tentative, which was a "nice" way of saying that truth, i.e., the recognition of reality, cannot exist. Knowledge of the three justifies empirical research, for it is based upon deep reflection of the world as disclosed by the senses. The pragmatic alternative is soon lost in the fog of context, eventually spouting contradictions as new discoveries. Logical positivism in all its variants, including that of Karl Popper, is a dull, leaden nihilism.

The human sensorium is not a veil of Maya, as supposed by Berkeley and sometimes Einstein.[299] J. Robert Oppenheimer, the principal developer of the American atom bomb, understood this: "There are children playing in the streets who could solve some of my top problems in physics, because they have modes of sensory perceptions that I lost long ago." [300]

Even the secondary perceptual characteristics, sound and color, do not prevent a scientific understanding of the frequencies which underlie them; rather, they assist it by picking out those which are most likely to affect life. Colors and sounds show which of those vibrations are harmful and which, beneficial.

And man is not the measure of all things, as the sophist Protagoras is supposed to have said, but a measurer of much, though not of the infinitesimal and the infinite. The following expressed thoughts of Confucius have been a mystery for thousands of years:

> "Yin, the visible, and Yang, the invisible, are called the Dao.
> The loving one discovers it and calls it love;
> The wise man discovers it and calls it wisdom."[301]

Although that was not the intention behind the writing of this book, the solution of this mystery is disclosed in its contents. The infinitesimal can never be experienced. Yet, it must exist. Imperceptibility is invisibility in the widest sense, but the finite world, the perceptible, is unintelligible without it.

But lest we enter into foolish pride, he also said:

> "That which cannot be fathomed in terms of Yin [the visible]
> and Yang [the invisible] is called God [Shen]."[302]

Confucius lived before Buddhism was introduced into China, which brought confusion to its culture.[303] According to him, the right way, the Dao, was on the verge of being lost altogether, even during his time.

Fine Western philosophy and the best of the ancient Chinese culture are not in contradiction. The world which is open to the senses and to instruments based upon them could be characterized as the Yin; the realm of the sub-finite and also the unending infinite, as the Yang. Mediating the two is the recognition of infinity as the immeasurable. Yet, knowing this only makes us more aware of our inherent limitations. It replaces dead positivism and prideful assertion with a lively humility. It is knowledge; and as such it deepens our understanding, but it provides no answers to human mortality.

For that, men and women must look elsewhere.

APPENDIX

(Cf., Chapter XVIII)

In Chapter XVIII, it was proved that the extra terms of the derivative are for the most part finite, and also that they are inexpungible. The theory presented here is fully in accord with this fact. It is up to the world to refute it, or at least to present an alternative account for the reality of the extra terms of the derivative.

Does non-mathematical evidence exist for it? One way would be to show how it gives a better account of some important phenomenon than does the received opinion. A case in point is in low temperature physics where the dominant school teaches that it is impossible to reach 0° Kelvin or Absolute Zero. They maintain that there exists a third law of thermodynamics, of which they have three formulations: (1) The unattainability form, which is that no process can reach $T = 0$ in a finite number of steps; (2) The absolute entropy form that the entropy goes to zero as $T \to 0$. (3) The entropy change form that this change goes to zero as $T \to 0$. Most advocates of the third law hold that the three expressions are equivalent.[304]

Against this stands the extra derivative physical theory, which holds that absolute zero is very difficult, but not impossible to obtain; that absolute zero is a state in which there can be no net transfer of heat energy from any body at that temperature to another; that this is the case, regardless of whether it is either a state in which there is no motion, or at the least, a minimum amount as determined by its parameters.

In reducing the temperature of an object, there is a withdrawal of heat energy. But even if there is a slight motion at 0°K, its energy would be in a final ground state; any force which would remove the latter would destroy the particle itself. Were the particle an element, it would break its atomic structure. In such a state, elasticity is great.

To continue with the extra-derivative theory: in the approach to Absolute Zero, energy is being extracted in tiny amounts. As the temperature descends by nano-Kelvins, there remains some elasticity. Since there is always some present during any temperature reduction, however, slight, a $\triangle E$ (energy) recovery persists at each and every step of that descent. This process is finite and not open to immeasurable subdivision. There has to be slightly more than enough force to reach the difference between whatever nano-Kelvin it is standing at just before the final descent to Absolute Zero and that final step; this extra force being necessary to overcome the elasticity barring the threshold of utter cold. To come short of that final withdrawal of the amount of energy required to reach that objective would be Sisyphean; the elastic forces would cancel the effort. And to get beyond it would damage the thing being cooled, disabling the elastic forces at zero degrees Kelvin and upsetting the integrity of the sample. What is required in order to reach the end of the defile is that exact of amount of energy withdrawal, just enough to reach the actual parametric ground state without causing damage.

This is best understood in contrast with a well-known opinion that there exists a certain "third law of thermodynamics." According this theory, as one gets close to absolute zero, the approach is asymptotic. One never reaches it. Yet, somehow, one can get on the other side—that there are negative degrees. These too are on an asymptotic approach to zero. And these negative degrees are supposed to be hot, not cold. Negative temperatures are imagined to be at their hottest just before they reach the allegedly impassible Absolute Zero. The difference between -.000000001 K and +.000000001K is supposed to be incredible heat and intense cold. In the words of Ralph Baierlein's *Thermal Physics*, a standard work: "All negative temperatures are hotter than all positive temperatures. Moreover the coldest temperatures are just above 0 K in the positive side, and the hottest temperatures are just below 0 K on the negative side."[305] So after one crosses over from the extremely cold positive side—however that is done—one immediately finds oneself into extreme heat. In fact, before crossing over from the negative side, it gets hotter and hotter and then—voila! Hot on the one side and utter coldness on the other side of the unreachable zero degree.

The extra-derivative physical theory is much simpler and does not involve the postulation of conceptually incongruent asymptotes. Under that theory, one would not get into an asymptotic process, unless one attempted to reach it by fractions of the preceding temperatures, instead of straightforward subtractions. Let us examine the two parts of the received theory separately, these being, first, the asymptotic structure of the approach to Absolute Zero and second, $\pm \infty$ temperatures.

The thermodynamical 3rd law opinion is made to seem plausible by their preferred equation for temperature: $1/T = (\partial S / \partial E)$, where T is absolute temperature, S is entropy, and E is energy.[306] In this formulation, when $T = 0°K$, the left side of the equation equals 1/0; since the inverse of zero is commonly held to be infinity, Absolute 0 could not be reached, unless the change in entropy were infinite, which is inexplicable.

If the idea that ∞ and 0 are inverses of each other be rejected, these theorists then conclude that Absolute Zero is left undefined, and the formula is simply an asymptote. Either way, that temperature is excluded.

But this formula cannot be universalized. In its terms, not only would it be impossible for T to be zero, but also that the change in energy (∂E) be zero. Inverting the formula to T = $\partial E/\ \partial S$, then neither can ∂S be zero. That these two must always be on the move is, perhaps, more than the advocates of that theory would want to assert. It is obvious that entropy can be left unchanged while temperature and energy change; their ratio could remain the same. The truth is that the formula is not universal; it cannot be applied in all contexts. A formulation appropriate to the derivative residual theory would be: $\partial T = \partial E/\ \partial S$. This allows for a final change to Absolute Zero equaling the final change in energy divided by the final change in entropy which brought about that situation. There is no problem with division by zero, no need to hypothecate an asymptote.

The advocates of the thermodynamical 3rd law theory use their formulation, because it allows them to define temperature in terms of entropy, in reverse of the classical formula. This reason for this preference is that they want to get away from the equilibration of temperature with average translational kinetic energy. Baierlein argues that the purpose of the temperature concept is not to "tell us about a physical system's amount of energy . . . ;" rather, it is "intended to tell us about a system's *hotness, its tendency to transfer* energy (by heating)."[307] By shifting the emphasis from kinetic energy to mere transfer, the incongruity of an infinite positive temperature standing but an impassible degree apart from infinite negative temperature is less apparent. And it is to such considerations we now turn.

Their opinion holds that the temperature in such a situation goes from -∞ to + ∞ and even that they are "physically equivalent." If they are, then what is the point of indicating them with opposite signs? This "transition" is impossible to represent on a Cartesian pair of axes. It must be supposed that without crossing zero, it can go from the positive infinite temperature to the negative. The justification is that the Boltzmann factor, $\exp(-E_j/kT)$ would be zero at minus infinite temperature, making the probability factor "perfectly flat . . . and so there is continuity in the probability distribution as the temperature makes the discontinuous switch from -∞ K to + ∞ K."[308] The statistical symbolism is supposed to supply the lacuna to be transited when the actual path of motion cannot be shown geometrically.

The received opinion also teaches that the "system passes smoothly from negative to positive temperature through the back door of infinite temperature."[309] But, under the formula, if the inverse of the infinite temperature be zero, then either the entropy change be zero or the energy change be infinite. But this last possibility, the theory forbids: According to Baierlein, one of the three conditions for a system to be capable of being at a negative temperature is that "the possible values of the system's energy must have a finite upper limit."[310] But this would be impossible, for an infinite temperature would require infinite energy.

Moreover, such a notion contradicts the very nature of an asymptote, according to which it would be impossible for the temperatures to ever be infinite. Instead, it would be a potential infinity, a process of the positive and negative temperatures getting hotter beyond human limits of foretelling, but always remaining finite nevertheless. Being of opposite sign, they are not equivalent and can never loop backwards. As Euclid proved long ago, N+1 is always finite. Being contradictory, the received theory is false on its own evidence.

It has been shown that the theory does not make sense with potential infinity. $\mathfrak{C}$ is out of the question. Since immeasurable infinity is only beyond human capacity, it would not work there, either. Such an infinity does not provide for a bypass of Absolute Zero. + and - would be veritable.

To return to an earlier point, temperature may not be adequately defined simply in terms of heat transfer. Average translational kinetic energy is an important part of it. But this difference between the two is tiny.[311] The truth is that this kinetic energy accounts for the overwhelming amount of the phenomenon of temperature. *Yet*, even if the heat transfer idea were correct, the extra-derivative theory is fully consistent with it, as was shown page 227.

The extra-derivative physical theory is based upon the idea that elasticity accounts for the fact that the extra terms of the derivative are ineradicable, that they, therefore, are as physical as the derivative itself, but that they do not show up as net physical effects. Unlike the received theory, this theory is consistent with itself as well as the data.

This theory also indicates that if the elasticity at Absolute Zero were broken in the instance of an atom, cold fission or cold fusion could result. Finding that would provide indirect evidence of the theory's truth.

Direct experimental evidence could be provided. That evidence would exist, if it were shown that after the temperature had reached Absolute Zero, there was at least one addition of heat energy to the sample which did not result in an increase in temperature. Proof would come if the body from which the energy was drawn registered no net change afterwards. Under this theory, heat capacity at Absolute zero would be zero, not an asymptotic approach to it. The addition, therefore, since it could not enter into the body nor raise the temperature of the patient nor lower that of the agent would have to have been rejected, thus revealing a threshold elasticity which must be crossed before there can be an addition in temperature. In that way, M. Lagrange's statement that the hidden result "should be measured by the effect which it would produce if it were not restrained from acting"[312] could be rendered intelligible to his critics.

It can be expected that the formula expressing it would show a change in energy and in entropy without any change in temperature. The term(s) that could produce this would be a genuine discovery.

BIBLIOGRAPHY

BOOKS

Adler, Irving, *A New Look At Geometry*, New York: The New American Library, 1966.

Archimedes, *The Works Of Archimedes*, Mineola, New York: Dover Publications, Inc., 2002, originally published in 1897 by Cambridge University Press.

Aristotle, *The Basic Works Of Aristotle*, Edited and with an introduction by Richard McKeon, New York: Random House, 1941.

Baierlein, Ralph, *Thermal Physics*, Cambridge, G.B., Cambridge University Press, 1999.

Ball, W. W. Rouse, *A Short Account of the History of Mathematics*, New York: Dover Publications, 1960, originally published in 1908.

Barr, Stephen, *Experiments In Topology*, New York: Thomas Y. Crowell Company, 1964.

Bell, Eric Temple, *Men Of Mathematics*, New York: Simon And Schuster, 1937.

Borschardt, Glenn, *The Ten Assumptions of Science: Toward a New Scientific Worldview*, New York: iUniverse, 2004.

Boyer, Carl, *The History Of The Calculus and Its Conceptual Development*, New York: Dover Publications, 1959, originally published in 1949 by Hafner Publishing Company.

Bunch, Bryan, *The Kingdom of Infinite Number: A Field Guide*, New York: W. H. Freeman and Company, 2000.

Burger, Edward B. and Starbird, Michael, *The Joy of Thinking: The Beauty and Power of Classical Mathematical Ideas*, Chantilly, VA, The Teaching Company, 2003.

Cajori, Florian, *A History Of Mathematical Notations, Two Volumes Bound As One*, New York, Dover Publications, 1993, originally published in 1923 by The Open Court Publishing Company.

Callahan, Jeremiah Joseph, *Euclid Or Einstein*, New York: Devin-Adair Company, 1931.

Cardano, Girolamo, *Ars Magna or The Rules of Algebra*, tr. Richard Witmer, New York: Dover Publications, Inc., 1993, originally published in 1968 by The MIT Press.

The Century Dictionary: An Encyclopedic Lexicon Of The English Language, Prepared under the Superintendence of William Dwight Whitney, PhD, LlD., Vol. IX, New York: The Century Co., 1911.

Courant, Richard, *Differential And Integral Calculus*, New York: Interscience Publishers, Inc., 1949.

Courant, Richard, and John, Fritz, *Introduction To Calculus and Analysis*, New York: Interscience Publishers, 1965.

Crow, Michael J, *A History Of Vector Analysis: The Evolution of the Idea of a Vectoral System*, New York: Dover Publications, 1985, originally published in 1967 by the University of Notre Dame Press.

Dauben, Joseph Warren, *Georg Cantor: His Mathematics and Philosophy of the Infinite*, Cambridge, MA: Harvard University Press, 1979.

Douglas, Jesse, *Survey of The Theory Of Integration*, New York: Scripta Mathematica, Yeshival College, 1941.

Eddington, Sir Arthur S., *The Nature Of The Physical World*, Cambridge, GB: Cambridge University Press, 1928.

Einstein, Albert, *The Meaning Of Relativity, 3rd E.*, revised including the *Generalized Theory Of Gravitation*, Princeton, New Jersey: Princeton University Press, 1950.

Einstein, Albert, *Relativity: The Special and the General Theory*, tr. Robert Lawson, New York: Crown Publishers, Inc., 1960.

Erickson, Peter, *The Stance Of Atlas: An Examination Of The Philosophy of Ayn Rand*, Portland, Oregon: Herakles Press, Inc., 1997.

Euclid, *The Thirteen Books Of Euclid's Elements, Translated from The Text Of Heiberg, With Introduction And Commentary by Sir Thomas L Heath, 2nd Ed., Vol. III*, New York: Dover Publications, 1956.

Euler, Leonard, *Introduction to Analysis of the Infinite, Book II*, tr. John Blanton, New York: Spring-Verlag, 1988.

Ferreirós, José, *Labyrinth of Thought: A History Of Set Theory and its Role in Modern Mathematics*, Boston, MA: Birkhäuser Verlag, 1999.

Flannery, Sarah, *In Code: A Mathematical Journey*, New York: Workman Publishing Co., 2001.

Gaukroger, Stephen, *Descartes An Intellectual Biography*, Oxford, G.B.: Clarendon Press, 1995.

Gödel, Kurt, *On Formally Undecidable Propositions of Principia Mathematica And Related Systems*, tr. B. Meltzer with Introduction by R. B. Braithwaite, New York: Basic Books, Inc., Publishers, 1962.

Greek Mathematics With An English Translation by Ivor Thomas, Cambridge, MA: Harvard University Press, 1993.

Hamilton, Sir William Rowan, *Lectures on Quaternions: Containing A Systematic Statement Of A New Mathematical Method: Of Which The Principles Were communicated In 1843 To The Royal Irish Academy, etc.*, Dublin: Hodges and Smith, 1853.

The Holy Bible Authorized King James Version, Comprising The Old And New Testaments With The Words Of Jesus Printed In Red And All Proper Names Self-Pronounced, Chicago: Spencer Press, 1947.

The Holy Bible In Modern English, Containing The Complete Sacred Scriptures Of The Old And New Testaments Translated Into English Direct From The Original Hebrew, Chaldee And Greek, tr. Ferrar Fenton, Merrimac, MA: Destiny Publishers, 1966, originally published in 1903.

Huntington, Edward V., *The Continuum And Other Types Of Serial Order*, 2nd *ed.*, Cambridge, MA: Harvard University Press, 1929.

Jesseph, Douglas M., *Berkeley's Philosophy Of Mathematics*, Chicago: The University of Chicago Press, 1993.

Jessuph, Douglas M., *Squaring the Circle: The War Between Hobbes and Wallis*: Chicago: The University of Chicago Press, 1999.

Joseph, H. W. B., *An Introduction To Logic*, Second Edition, Revised, Oxford, At The Clarendon Press, 1916.

Kant, Immanuel, *Immanuel Kant's Critique Of Pure Reason*, tr. F. Max Müller, London: The Macmillan Company, 1896.

Kline, Morris, *Calculus: An Intuitive and Physical Approach*, 2 volumes, New York: John Wiley and Sons, Inc., 1967.

Kline, Morris, *Mathematical Thought from Ancient to Modern Times, Volume 3*, New York: Oxford University Press, 1990—paperback.

Kline, Morris, *Mathematics: The Loss Of Certainty*, New York: Oxford University Press, 1980.

Kraus, Gerhard, *Has Hawking Erred?* London: James Publishing Co., 1993.

Lieber, Lillian R., *Infinity, with drawings by Hugh G. Lieber*, New York: Holt, Rinehart and Winston, 1953.

Lieber, Lillian R.with drawings by Hugh Gray Lieber, *Non-Euclidean Geometry or Three Moons in Mathesis*, New York: Academy Press, 1931.

Lebesgue, Henri, *Measure And The Integral, edited with a biographical essay by Kenneth O. May*, San Francisco, Holden-day, Inc., 1966.

Maclaurin, Collin, *A Treatise on Fluxions, Vol. 1*, William Baynes And William Davis, 1801.

Maor, Eli, *e The Story Of A Number*, Princeton, New Jersey: Princeton University, 1994.

Maxwell, James Clerk, *Matter And Motion Notes And Appendices By Sir Joseph Larmor*, New York: Dover Publications, 1991, originally published in 1920 by the Society For Promoting Christian Knowledge.

Mazur, Barry, *Imaging Numbers (particularly the square root of minus fifteen)*, New York: Farrar Straus Giroux, 2003.

Mill James, *Analysis Of The Phenomena Of The Human Mind, Ed. With Additional Notes By John Stuart Mill*, 2 Volumes, London: Longmans, Green, Reader, and Dyer, 1878.

Mill, John Stuart, *An Examination of Sir William Hamilton's Philosophy and of The Principal Philosophical Questions Discussed in his Writings*, ed. J. M. Robson, Toronto, Canada: University Of Toronto Press, 1979.

Moe, Russel, *Polyscience and Christianity*, Kearney, NE: Morris Publishing, 2004.

Mott-Smith, Morton, *Principles Of Mechanics Simply Explained, Revised Ed.*, New York: Dover Publications, 1963, originally published in 1931 by Appleton and Company.

Nahim, Paul J., *An Imaginary Tale: The Story of $\sqrt{-1}$*, Princeton, New Jersey: Princeton University Press, 1998.

Newton, Sir Isaac, *Mathematical Principles of Natural Philosophy And His System Of The World, Volume One: The Motion Of Bodies*, tr. by Andrew Motte in 1729; the translation revised and supplied with an appendix by Florian Cajori, Berkeley, CA: University Of California Press, 1966.

Olmstead, John M., *The Real Number System*, New York: Appleton-Century Crofts, New York, 1962.

Pesic, Peter, *Abel's Proof: An Essay on the Sources and Meaning of Mathematical Unsolvability*, Cambridge, MA: The MIT Press, 2003.

Rand, Ayn, *Introduction To Objectivist Epistemology*, With an Additional Essay by Leonard Peikoff, ed. by Harry Binswanger and Leonard Peikoff, expanded second edition, New York: Meridian, 1988.

Ray, Joseph, *Ray's New Higher Arithmetic: A Revised Edition Of The Higher Arithmetic*, Cincinnati: van Antwerp Bragg & Co., 1880.

Robinson, Abraham, *Non-Standard Analysis*, Princeton, New Jersey: Princeton University Press, 1966.

Rosenblatt, Judah and Bell, Stoughton, *Mathematical Analysis for Modeling*, Boca Raton, FL: CRC Press, 1999.

Russell, Bertrand, *The Principles Of Mathematics*, 2nd Ed., paperback, New York: W. W. Norton & Co., 1996.

Sawyer, W.W, *Prelude to Mathematics*, Middlesex, G.B.: Penguin Books, 1955.

Smith, Crosbie, and Wise, M. Norton, *Energy and Empire: A biographical study of Lord Kelvin*, Cambridge: Cambridge University Press, 1989.

Spengler, Oswald, *The Decline Of The West, Volume I: Form and Actuality*, tr. Charles Francis Atkinson, New York: Alfred A. Knopf, 1926.

Starbird, Michael, *Change and Motion: Calculus Made Clear, Part II*, Chantilly, VA: The Teaching Company, 2001.

Starbird, Michael, *The Joy of Thinking: The Beauty and Power of Classical Mathematical Ideas*, Lecture One, Chantilly, VA: The Teaching Company, 2003.

Turner, Dean and Hazelett, Richard, *The Einstein Myth and the Ives Papers: A Counter-Revolution in Physics*, Old Greenwich, Connecticut: The Devin-Adair Company, 1979.

The Universal Encyclopedia of Mathematics, Foreword by James R. Newman, New York, New York: Mentor Books, 1965, originally Published in German by Bibliographisches Institut in Mannheim, 1960.

Waismann, Friedrich, *Introduction To Mathematical Thinking: The Formulation Of Concepts In Modern Mathematics*, New York: Harper & Row, 1959.

Welch, Robert, *The Romance Of Education*, Belmont, MA: Western Islands, 1973.

Wilson, Edwin Bidwell, *Advanced Calculus: A Text Upon Select Parts of Differential Calculus, Differential Functions, Integral Calculus, Theory Of Functions, With Numerous Exercises*, Boston: Ginn And Company, 1912.

Wilson John Cook, *Statement And Inference with Other Philosophical Papers*, 2 Volumes, ed. by A.S.L. Farquarson, Oxford: At The Clarendon Press, 1926.

ARTICLES

Apostol, Tom A., "A Visual Approach to Calculus Problems," *Engineering & Science, Vol. LXIII, Number 3,* 2000.

Bergman, David L., "Observations of the Properties of Physical Entities Part I—Nature of the Physical World," *Foundations Of Science, Vol. 7, no. 1,* February 2004.

—"Hypertension treatment proves 88 percent effect in Brazilian Study," *HSI Heath Sciences Institute,*—

Hymowitz, Carol, "Home Depot's CEO Led a Revolution, But Left Some Behind,' *The Wall Street Journal*, Tuesday, March 16, 2004.

Lindner, Henry H., "Beyond Consciousness to Cosmos—Beyond Relativity and Quantum Theory to Cosmic Theory," *Physics Essays*, Volume 15, Number 1, 2005.

Lucas, Joseph and Lucas, Charles W. Jr., "A Physical Model for Atoms and Nuclei—Part I," *Foundations of Science, Volume 5, Number 1*, February, 2002.

Nelson, J. J., "Some Experimental Incoherencies of Riemannian Space," *Philosophical Mathematica*, 12 (1975), pp. 66-75; reproduced by Dean Turner and Richard Hazelett, *The Einstein Myth and the Ives Papers: A Counter-Revolution in Physics*, Old Greenwich, Connecticut, 1979.

Pisaturo, Ronald and Marcus, Glenn D., "The Foundation of Mathematics," *The Intellectual Activist, Vol. 8 no. 4*, July 1994, and Vol. 8, No. 5, September 1994.

INTERNET

Jesseph, Douglas M., "Leibnitz on the Foundations of the Calculus: The Question of the Reality of Infinitesimal Magnitudes," *http://muse.jhu.edu/demo/posc/6.1jesseph.html* (04/04/2002.) Oppenheimer, J. Robert, *http://www.brainyquote.com/quotes/quotes/j/jrobertop.106068.html* (06/14/2005.) *Http://ingeb.org/songs/forwanto.html* (06/14/2005.)

COVER PHOTOGRAPHS

"Gaseous Pillars in M16, Eagle Nebula, Hubble Space Telescope, WFPC2," Greenbelt, MD/Goddard Space Flight Center.

Dot pattern, in Hornung, Clarence, *Background Patterns, Textures And Tints*, New York: Dover Publications, 1976.

ENDNOTES

1. Carol Hymowitz, "Home Depot's CEO Led a Revolution, But Left Some Behind," *The Wall Street Journal*, Tuesday, March 16, 2004, p. B1

2. Morris Kline, *Mathematics: The Loss of Certainty*, (New York: Oxford University Press, 1980), Preface.

3. Ayn Rand, *Introduction To Objectiviist Epistemology With an Additional Essay by Leonard Peikoff*, ed. By Harry Binswanger and Leonard Peikoff, expanded second edition, (New York: Meridian, 1988), p. 36.

4. *Ibid.*, p. 262

5. *Ibid.*, p. 266.

6. Dominquez José Ferreirós, *Labyrinth of Thought: A History of Set Theory and its Role in Modern Mathematics*, (Boston, MA: Birkhäuser Verlag, 1999), p. 58.

7. Sir Arthur S. Eddington, *The Nature Of The Physical World*, (Cambridge, Great Britain: Cambridge University Press, 1928), p 332.

8. *Ibid.*, p. 308.

9. Aristotle, *Categories*, 4b 20.

10. Richard Dedekind, quoted in Ferreirós, *op. cit.*, p. 117.

11. Peter Erickson, *The Stance Of Atlas: An Examination Of The Philosophy of Ayn Rand*, (Portland, Oregon: Herakles Press, Inc.,1997), p. 87.

12. *The Thirteen Books Of Euclid's Elements, Translated from The Text Of Heiberg, With Introduction And Commentary by Sir Thomas L. Heath, 2nd Ed.*, Vol. 111, (New York: Dover Publications, 1956), pp. 1-254, *passim.* Cf., especially, Sir Thomas' notes.

13. Henri Lebesgue, *Measure And The Integral, edited with a biographical essay by Kenneth O. May*, (San Francisco: Holden-day, Inc, 1966), p. 129.

14. *Loc. cit.*

15. *Ibid.*, p. 128.

16. *Ibid.*, p. 94.

17. Byran Bunch, *The Kingdom of Infinite Number: A Field Guide*, (New York: W. H. Freeman and Company, 2000), p. 335.

18. Ayn Rand, *Introduction to Objectivist Epistemology With an Additional essay by Leonard Peikoff*: expanded Second Edition, ed by Harry Binswanger and Leonard Peikoff (New York: Penguin Company, 1990), p. 18.

19. Ronald Pisaturo and Glenn D. Marcus, "The Foundation of Mathematics," *The Intellectual Activist, Vol. 8, No. 4*, July 1994, p. 4.

20. *Ibid.*, p. 7-8.

21. *Ibid.*, p. 11.

22. *Ibid.*, p. 8.

23. *Ibid.*, p. 7.

24. *Ibid.*, pp. 4-6.

25. Ayn Rand, *op. cit.*, p. 7.

26. Pisaturo and Marcus, *op. cit., Part II, Vo. 8*, September 1994, p. 8.

27. *Loc. cit.*, p. 8.

28. Sir William Rowan Hamilton, *Lectures on Quaternions: Containing A Systematic Statement Of A New Mathematical Method; Of Which The Principles Were Communicated In 1843 To The Royal Irish Academy*, etc., (Dublin: Hodges and Smith, 1853), pp. 5-6.

29. Pisaturo and Marcus, *op. cit.*, p. 8.

30. *Loc. cit.*

31. Pisaturo and Marcus, *II*, p. 10.

32. *Loc. cit.*

33. Aristotle, *Physics 207a 8-10*, from *The Basic Works Of Aristotle*, tr. Richard McKeon, (New York: Random House, 1941), p. 206.

34. Ayn Rand, *op. cit.*, p. 18.

35. This is the case even for some finite terms. As the logician H. W. B. Joseph writes: "In defining a privative or negative concept it is inevitable. A bachelor is an unmarried man; and the very meaning of the term is to deny the married state." *An Introduction To Logic, 2nd ed., revised*, (Oxford At The Clarenden Press, reprinted 1966), p. 114.

36. Sarah Flannery, *In Code: A Mathematical Journey*, (New York: Workman Publishing Co., 2001), p. 69.

37. Euclid, *The Thirteen Books Of Euclid's Elements, With Introduction and Commentary by Sir Thomas Heath, Volume I*, (New York: Dover Publications, 1956), p.153.

38. Friedrich Waismann, *Introduction To Mathematical Thinking*, (New York: Harper & Row, 1951), p. 73.

39. Douglas M. Jesseph, *Berkeley's Philosophy Of Mathematics*, (Chicago: University of Chicago: 1993), pp. 57-62.

40. Bunch, *op. cit.*, p. 337.

41. Stephen Barr, *Experiments In Topology*, (New York: Thomas Y. Crowell Company, 1964), p. 150.

42. Bertrand Russell, *The Principles Of Mathematics*, (New York: W. W. Norton, reissued 1996), p. 408.

43. Pisaturo and Marcus, *op. cit., II*, p. 12.

44. *Ibid. II*, p. 15.

45. Ferreirós, *op. cit.*, p. 55.

46. *Ibid.*, p. 103.

47. *loc. cit.*

48. *Ibid.*, p. 104.

49. *Ibid.*, p. 119.

50. Morris Kline, *Mathematical Thought from Ancient to Modern Times, Volume 3*, (New York: Oxford University Press, 1990—paperback), p. 951.

51. *Ibid.*, p. 951.

52. Edward V. Huntington, *The Continuum And Other Types Of Serial Order*, 2nd ed., (Cambridge, MA., 1929), p. 13.

53. *Ibid.*, p. 135.

54. Kline, *op. cit.*, p. 1033.

55. Carl Boyer, *The History Of The Calculus and Its Conceptual Development*, (New York: Dover Publications, 1959), p. 293.

56. Leonard Peikoff, "The Analytic-Synthetic Dichotomy," Rand, *op. cit.*, p. 116.

57. If someone were really afraid of eventually reaching a discontinuity in space thereby preventing them from accepting a return to a number line extending indefinitely in space, then let them stay with the region of space they feel has been "empirically" proven and simply cut their number line in half when they have to include greater numbers. If one supposes, as most of them do, that any interval can be infinitely divided, they need not make their line any greater. If they are afraid that the planet might move into such a region, well then, how can they trust their free constructs?

58. Huntington, *op. cit.*, p. 19. Huntington's other postulate is: "*(Dedekind's postulate)* If K$_1$ and K$_2$ are any two non-empty parts of K, such that every element of K belongs either to K$_1$ or to K$_2$ and every element of K$_1$ precedes every element of K$_2$, then there is at least one element X in K such that: (1) any element that precedes X belongs to K$_1$, and (2) any element that follows X belongs to K$_2$.*"

59. *Ibid.*, p. 21.

60. *Ibid.*, p. 44.

61. Ayn Rand, *op. cit*, pp. 58-9.

62. Ferreirós, op. *cit.*, p. 191.

63. For reasons which will be presented elsewhere, this writer does not even believe the idea of an infinite sequence of digits is appropriate for irrational numbers; but their problem there is not as serious as it is in the case of what is being discussed in this paper. With the irrational numbers, one can, at least, approximate.

64. *Ibid.*, p. 136.

65. Euclid, *Book I, (Heath, Vol I), op. cit.*, p. 153.
66. Pisaturo and Marcus, *op. cit. II*, pp. 14-15.
67. *Ibid.*, p. 14.
68. Aristotle, "On The Soul, 409a, 4-5", *op. cit.*, p. 549.
69. Kline, *op. cit.*, p. 1011.
70. Heath, *Euclid's Elements I, op. cit.*, pp. 193, 232-3.
71. Richard Dedekind, quoted in Carl B. Boyer, *The History Of The Calculus And Its Conceptual Development*, tr. Richard Courant, (New York: Dover Publications, 1959), p. 296. "A system S is said to be *infinite* when it is similar to a proper part of itself; in the contrary case S is said to be a finite system."
72. Carl B. Boyer, *The History of the Calculus and its Conceptual Development, Forward by Richard Courant*, (New York: Dover Publications, 1959), p. 296.
73. *Ibid.*, p. 270.
74. Huntington, *op. cit.*, p. 7.
75. Lillian R. Lieber, *Infinity*, with drawings by Hugh G. Lieber, (New York: Holt, Rinehart and Winston, 1953), p. 142-143.
76. Kline, *Mathematics: The Loss of Certainty*, p. 200. "He [Cantor] also gave the argument that the irrational numbers, such as $\sqrt{2}$, when expressed as decimals involved actually infinite sets because any finite decimal could only be an approximation."
77. Huntington, *op. cit.*, p. 23.
78. *Loc. cit.*
79. Georg Cantor, quoted in Ferreriós, *op. cit.*, p. 275.
80. *Ibid.*, p. 279.
81. Lieber, *op. cit.*, p. 94.
82. *Ibid.*, p. 96.
83. Eli Maor, *op. cit.*, p. 215.
84. *Loc. cit.*
85. Rand, *op. cit.*, pp. 72-73. "The optional area includes also the favorite category (and straw man) of modern philosophers: the 'Borderline Case'. By 'Borderline Case,' they mean existents which share some characteristics with the referents of a given concept, but lack others; or which share some characteristics with the referents of two different concepts . . . ."
86. Lillian R. Lieber with drawings by Hugh Gray Lieber, *Non-Euclidean Geometry or Three Moons in Mathesis*, (New York: Academy Press, 1931), p. 4.
87. *Euclid I, op. cit.*, p. 153.
88. Kline, *Mathematics: The Loss Of Certainty, op. cit.*, p. 191.
89. Morris Kline, *Mathematical Thought From Ancient To Modern Times*, Vol 3, *op. cit.*, p. 1011.
90. *Loc. cit.*
91. Kline, *Mathematical Thought, etc., op. cit.*, p. 1013

92. Ferreirós, *op. cit.*, p. 102.

93. John Cook Wilson, *Statement And Inference with Other Philosophical Papers. Vol. II*, ed. by A.S.L. Farquharson (Oxford: At The Clarendon Press, 1926), p. 465.

94. Archimedes, "The Method of Archimedes Treating of Mechanical Problems—To Erastothenes," *The Works Of Archimedes*, (Mineola, New York: Dover Publications, Inc., 2002), p. 17.

95. Kline, *Mathematics: The Loss of Certainty*, p. 180.

96. *Loc. cit.*

97. Proclus, quoted by Sir Thomas L. Heath in *The Works Of Archimedes*, ed. by T.L. Heath, (Mineola, New York: Dover Publications, 2002), p. 4.

98. Bertrand Russell, *The Principles Of Mathematics*, 2nd Ed., paperback, (New York: W. W. Norton & Co., 1996.), p.410.

99. Cf., Ayn Rand, *op. cit,*. pp. 40-41.

100. Bertrand Russell, *op. cit.*, p. 408.

101. John Cook Wilson, *op. cit.*, p. 711.

102. Immanuel Kant, *Immanuel Kant's Critique Of Pure Reason*, tr. F. Max Müller, (London: The Macmillan Company, 1896), p. 353.

103. John Cook Wilson, *op. cit.*, p. 564.

104. Glenn Borchardt, *The Ten Assumptions of* Science, (New York: i Universe, Inc. 2004, pp, 89-93.

105. Douglas M. Jesseph, *Squaring the Circle: The War between Hobbes and Wallis.* (Chicago: The University of Chicago Press, 1999), p. 170. This figure was originally drawn by Thomas Hobbes in a controversy he had with John Wallis. Hobbes' point, however, in all probability, was very different from mine. He may not even have been aware of the figure's ambiguity. He did not accept infinitesimals.

106. A way to draw this figure without crossing any point twice would be as follows: (a) Begin at the right and draw the line segment toward the left, then stop. (b) From that point draw the upper loop counterclockwise, completing it just to the left of the very point from which the circular construction began and the line segment stopped. (c) Starting just below the line segment, draw the lower loop clockwise, stopping just to the left of the point at which the upper loop ended. (d) Then draw the remaining line segment plump with the line segment already drawn. This figure would be continuous; there was no breach from one point to the next (even in the Weirstrassian sense).

107. Edward V. Huntington, *op. cit.*, p. 34.

108. Michael J. Crowe, *A History Of Vector Analysis: the Evolution of the Idea of a Vectoral System*, (New York: Dover Publications, 1985), p. 228.

109. Edwin Bidwell Wilson, *Advanced Calculus: A Text Upon Select Parts of Differential Calculus, Differential equations, Integral Calculus, Theory of Functions, with Numerous Exercises*, (Boston: Ginn and Company, 1912), p. 63.

110. *Loc. cit.*

111. *Loc. cit.*
112. Kline, *Mathematics: The Loss of Certainty*, *op. cit.*, pp. 274-275.
113. Abraham Robinson, *Non-Standard Analysis*, (Princeton, New Jersey: Princeton University Press, 1996), p.56. Cf., Chapters XVI and IXX.
114. Kline III, *op. cit.*, pp. 1018-19.
115. Joseph Warren Dauben, *Georg Cantor: His Mathematics and Philosophy of the Infinite*, (Cambridge, MA: Harvard University Press, 1979), p. 294.
116. *Ibid.*, p. 130.
117. *Loc. cit.*
118. There is more than one kind of irrational magnitude. The common type is like $\sqrt{2}$. Let any example of these be called a *"common irrational."* But there is another kind which is not commensurate with any of these. Technically, they comprise all those irrational numbers that are not solutions for algebraic equations. They are called *"transcendental numbers."* Examples are π and e.
119. Ferreriós, *op. cit.*, p. 291.
120. Sometimes it is objected that "pure space has been found nowhere. Experimentally, the closest we have ever come to empty space is a partial vacuum."[Borchardt, *op. cit.*, p. 59]. The answer is that it has not been detected, because it cannot be found through sense data. It is a discovery of reason. The important physical concept of density presupposes it. How can a balloon of a given density increase in size, unless there is some place for it to expand? The phenomenon of entropy also requires it.
121. *Ibid.*, p. 108.
122. Russel Moe, *Polyscience and Christianity*, (Kearney, NE: Morris Publishing, 2004), p. 155. The quotation is exact. Mr. Moe had spelled "traveled" as "travelled," an infrequent spelling still found in a dictionary.
123. Morris Kline, *Mathematics: The Loss of Certainty*, p. 86.
124. E.T. Bell, *Men Of Mathematics*, (New York, N.Y.: Simon And Schuster, 1937), p. 491.
125. John O. Nelson, "Some Experimental Incoherencies of Riemannian Space," *Philosophical Mathematica 12 (1975)*: 66-75; reproduced in Dean Turner and Richard Hazelett, *The Einstein Myth and the Ives Papers: A Counter-Revolution in Physics*, (Old Greenswich, Connecticut, 1979), p. 227.
126. J. J. Callahan, *Euclid Or Einstein*, (New York: Devin Adair Company, 1931), p. 184.
127. Turner and Hazelett, *op. cit.*, p. 30.
128. Ray, *op. cit.*, p. 379.
129. Lieber, *Infinity*, p. 110.
130. Adler, Irving, *A New Look At Geometry*, (New York, NY: The New American Library, 1966), p. 84.
131. ". . . there is no temporal "before" and "after" in the *logical* structure of mathematics." Barry Mazur, *Imaging Numbers (particularly the square root of minus fifteen)*, (New York: Farrar Straus Giroux, 2003), p. 150.

132. Adler, *op. cit.*, p. 271.
133. David Harriman, quoted by David L. Bergman, "Observations of the Properties of Physical Entities Part 1—Nature of the Physical World," *Foundations Of Science*, vol. 7, no. 1 (February 2004), p. 5.
134. Joseph Ray, *op. cit.*, P. 43.
135. *Loc. cit.*
136. With respect to a purely abstract equation—one that connects neither to what the drawing is supposed to show, nor to physical reality, however, someone might argue that this division is impossible, since it does not represent anything beyond itself. But if it is outside of reality, how can it be understood, since the ultimate reference is to infinitesimals, whether acknowledged or not? On the other hand, if the equation is just an illustration for a student or a general form, as in F = ma, then its use is reasonable.
137. Richard Courant and Fritz John, *Introduction To Calculus and Analysis*, (New York: Interscience Publishers, 1965), p. 166-8.
138. H.W.B. Joseph, *op. cit.*, p. 111.
139. *Ibid.*, p. 112.
140. *Loc. cit.*
141. *Loc. cit.*
142. *Ibid.*, p. 113.
143. *Ibid.*, p. 114.
144. Joseph Lucas and Charles W. Lucas, Jr., "A Physical Model for Atoms and Nuclei—Part I," *Foundations of Science*, Volume 5, Number 1, (February 2002), p. 1 ". . . electron scattering experiments, for which Robert Hofstadter received the Nobel Prize in 1961, have shown that neutrons, protons, and other elementary particles have a measurable finite size, an internal charge distribution (indicative of internal structure), and elastically deform in interactions. The size and shape of the electron was measured by Compton and refined more completely by Bostick, his last graduate student."
145. Edwin Wilson, *op. cit.*, p. 62.
146. Ayn Rand, *Op. cit.*, p. 28. "The process of observing facts of reality and of integrating them into concepts is, in essence, a process of induction. The process of subsuming new instances under a known concept is, in essence, a process of deduction."
147. Erickson, *op. cit.*, pp.79.
148. Euclid, *op. cit.*, p. 155.
149. Erickson, *op. cit.*, pp. 122-123.
150. Peter Pesic, *Abel's Proof: An Essay on the Sources and Meaning of Mathematical Unsolvability*, (Cambridge, MA: The MIT Press, 2003), p. 65.
151. *Ibid.*, pp. 138-139.
152. Henri Lebesgue, *op. cit.*, p. 179.
153. W.W. Rouse Ball, *A Short Account of the History of Mathematics*, (New York: Dover Publications, 1960), p. 224.

154. Oswald Spengler, *The Decline Of The West, Volume I: Form and Actuality*, tr. Charles Francis Atkinson (New York: Alfred A. Knopf, 1926), pp. 77, 89-90.

155. Albert Einstein, *The Meaning Of Relativity 3rd Ed., revised including the Generalized Theory Of Gravitation*, (Princeton, New Jersey: Princeton University Press, 1950), pp. 33-4; 37-38, 43, 84-85, 87, 98.

156. Paul J. Nahin, *An Imaginary Tale: The Story Of $\sqrt{-1}$*, (Princeton, New Jersey: Princeton University Press, 1998), p. 83.

157. *Ibid.*, pp. 48-54.

158. Bryan Bunch, *The Kingdom of Infinite Number: A Field Guide*, (New York: W.H Freeman and Company, 2000), p. 310.

159. Lebesgue, op. *cit.*, p. 94.

160. Sir William Hamilton, *op. cit.*, pp. 43-44.

161. Morris Kline, *Mathematics: The Loss Of Certainty*, p. 156.

162. W. W. Sawyer, *Mathematician's Delight*, Middlesex, (England: Penguin Books, 1959), p. 219.

163. Karl Friedrich Gauss, quoted by Nahin, *op. cit.*, p. 82. The actual words were "vera umbrae umbra."

164. Kline, *op. cit.*, pp. 118-19.

165. Eli Maor, *e: The Story Of A Number*, (Princeton, New Jersey: Princeton University, 1994), p. 177.

166. Since, in the veritable system, y is defined for negative values of x, that is no objection to extending the equation for negative values.

Let us next consider the problem of a coordinate system appropriate for veritable numbers. In the standard four-quadrant Cartesian system, a pair of crossed veritable axes could produce figures in only the 1st and 3rd quadrants; this is because the 2nd and 4th are both negative times positive. Instead, let us eliminate these two unworkable axes, leaving the two others. Then, let us take the third quadrant (negative times negative) and using the 0 point as a hinge, rotate it counter-clockwise until it lies under the first quadrant (positive times positive); then orient the combination as one wills. Then let us restrict any figures one might draw so that no point in them crosses the double zero line.

Next, let us erase the unnecessary neighboring horizontal line and designate the remaining one as the zero line with positive on one side and negative on the other.

Note that each quadrant is either fully positive and negative. It is well-suited to the veritable system. An equation like $f(x) = |x|$, could be represented therein, with a lower derivative of -1 and a upper derivative of +1.

On one side of the horizontal veritable axis, every value is + and on the other side, every value is -. Immediately above and below the zero line are those positionals, which are beneath the level of the finite

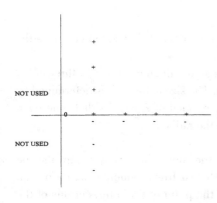

When the figure is going to be in the 1ˢᵗ and 3ʳᵈ axes only, then one may revert to the standard coordinate arrangement. Figures which normally straddle three or more coordinates would have to have their axes shifted. For example, the circle would require as its center some other position than zero. There are certain situations where the veritable is to be preferred over the real.

At present, the author is exploring a non-Cartesian idea which will provide a much greater range of representation for the veritable.

167. This is distinguished from the term "zero absolute" used in the 9ᵗʰ chapter, which was shown to be very different.

168. An example of such a contradiction is an attempt early in the last century by Edward Kasner to work with an axis which is complex, a contradiction, since two incompatible numbers are added together and then used for an axis. Even if a series of complex points in a single quadrant, or across two, should form a straight line, the placement of zero alone would require a re-calibration.: If imaginary or real numbers of zero value were used, it would be no different from using a pure imaginary or real line. Kasner applied his confusion to the problem of calculating a chord length and came forth with the impossible conclusion that the arc length was 6% shorter than the straight line. Instead of concluding that he must have made an error, both he and the source listed considered it a new discovery. (Cf., Nahin, *op. cit.*, pp. 106-107).

169. Below the level of the unit circle, however, it is possible to produce a linear square: for instance, $[\{\sqrt{-1}\}^{1/2}]^2 = \sqrt{-1}$. But that is another subject.

170. "0" may not be quite the same in the veritable and real system, but the same point can be used for two different purposes, providing that there is no admixture, as discussed above.

171. Richard Courant and Fritz John, *Introduction To Calculus And Analysis*, (New York: Interscience Publishers, 1965), p. 168.

172. Leonard Euler, *Introduction to Analysis of the Infinite, Book II*, tr. John Blanton, (New York: Spring-Verlag, 1988), p. 337.

173. D.F. Gregory, "On Impossible Logarithms," cited in Crosbie Smith and M. Norton Wise, *Energy and Empire: A biographical study of Lord Kelvin*, (Cambridge: Cambridge University Press, 1989), p. 185

174. John M. Olmstead, *The Real Number System*, (New York: Appleton-Century Crofts, New York, 1962). In this work, I have not found any treatment of this question.

175. Bryan Bunch, *op. cit.*, p. 190.

176. Ray, *op. cit.*, p. 175.

177. Euclid, Book V, def. 2.
178. The only way to make that reconcile would be to use $\sqrt{-5}$, instead of $\sqrt{-1}$ as the unit for i.
179. A rotational unit is here defined as motion across the outer rim of a three sided sector with the center of motion at the apex. The size of the unit is determined by the number of sectors in which a complete rotation of $2\pi r$ is divided, where r = radius. The more common measurement is the radian.
180. *Ibid.*, p. 4.
181. Against this, it might be asserted that the ambiguous figure can also be regarded as a roundish figure eight with two line segments beginning on each side just to one side or the other of the point of the intersections of the loops on the 8—in which case, an intersecting point of tangency could be regarded only as part of a construction line and not part of the figure. But this does not alter the main argument, which is that the definition of a tangent is of all importance.
182. Judah Rosenblatt and Stoughton Bell, *Mathematical Analysis for Modeling*, (Boca Raton, FL: CRC Press, 1999), p. 68.
183. *The University Encyclopedia Of Mathematics*, Forward by James R. Newman, (New York: Mentor, 1965), p. 424.
184. *The Century Dictionary: An Encyclopedic Lexicon Of The English Language, Prepared under the Superintendence of William Dwight Whitney, PhD., LLD. Vol IX*, (New York: The Century Co., 1911), pp. 6178-9.
185. Euclid *Book III, Proposition 16, op. cit., Vol. 2*, pp. 37-38.
186. T. Heath, "Introduction—The Terminology Of Archimedes," Archimedes, *op. cit.*, p. clxxiii.
187. Sir Isaac Newton, *Mathematical Principles of Natural Philosophy And His System Of the World, Volume One: The Motion Of Bodies*, tr. by Andrew Motte in 1729; the translation revised and supplied with an appendix by Florian Cajori, (Berkeley, CA: University Of California Press, 1966), p. 32.
188. Colin Maclaurin, *A Treatise On Fluxions, Vol. 1* (William Baynes And William Davis, 1801), p. 179.
189. Waismann, *op. cit.*, p. 148.
190. Maclaurin, *op. cit.*, p. 179.
191. *Loc. cit.*
192. *Ibid,* p. 230.
193. Edwin Wilson, *op. cit.*, p. 9.
194. Maclaurin I, *op. cit.*, p. 242.
195. R. Courant, *Differential And Integral Calculus*, (New York: Interscience Publishers, Inc., 1949), p. 163.
196. Boyer, *op. cit.*, p. 227.
197. Augustine Louis Cauchy, quoted in Boyer, *op. cit.*, p. 272.

198. Boyer, *op. cit.*, p. 231.

199. Michael Starbird, *Change and Motion: Calculus Made Clear. Part II*, (Chantilly, VA: The Teaching Company, 2001) p. 19.

200. *Loc. cit.*

201. Morris Kline, *Calculus: An Intuitive and Physical Approach, Part Two*, (New York: John Wiley and Sons, Inc. 1967), p. 332.

202. Kline III, *op. cit.*, p. 952. A predecessor was the Frenchman Cauchy. As Karl Boyer comments: "Cauchy's theorem—that a necessary and sufficient condition that the sequence converge to a limit is that the difference between S_p and S_q for any values of p and q greater than n can be made less in absolute value than any assignable quantity by taking n to converge within itself. The necessity of the condition follows immediately from the definition of convergence, but the proof of the sufficiency of the condition requires a previous definition of the system of real numbers, of which the supposed limit S is one." Boyer, *op. cit.*, p. 281. But, in truth, this sum is not reached by lower numbers.

203. Waismann, *op. cit.*, p. 151.

204. *Ibid.*, pp. 151-2.

205. Huntington, *op. cit.*, p. 65.

206. Boyer, *op. cit.*, p. 290.

207. Karl Boyer, *op. cit.*, p. 196.

208. Florian Cajori, *A History Of Mathematical Notations, Two Volumes Bound As One, II*, (Dover Publications: New York, 1993), p. 218.

209. Waismann, *op. cit.*, p. 151. Consider this statement by Edwin Wilson, *op. cit.* p. 65: According to him, dy/dx = f'(x) "is the quotient of two finite quantities of which dx may be assigned at pleasure. This is true if x is the independent variable." From this he gets the following theorem which he regards as important: *"The quotient dy/dx is the derivative of y with respect to x no matter what the independent variable may be.* It is this theorem which really justifies writing the derivative as a fraction and treating the component differentials according to the rules of ordinary fractions."

210. Edwin Wilson, *op. cit.*, p. 67.

211. Colin Maclaurin, *op. cit., I*, p. 242.

212. Abraham Robinson, *op. cit.*, p. 68.

213. Morris Kline, *Calculus: An Intuitive and Physical Approach*, Vol. 1, p. 236.

214. *Ibid.*, pp. 236-7.

215. Tom A. Apostol, *"A Visual Approach to Calculus Problems"*, *Engineering & Science, (Vol LXIII, Number 3, 2000)*, pp. 23, 31.

216. Morton Mott-Smith, *Principles Of Mechanics Simply Explained, Revised Ed.*, (New York: Dover Publications, 1963), p. 90.

217. *Ibid.*, p. 90-91.

218. James Clerk Maxwell, *Matter And Motion, Notes and Appendices By Sir Joseph Larmor*, (New York: Dover Publications, 1991), p. 130.

219. *Ibid.*, p. 2. "When a material system is considered with respect to the relative position of its parts, the assemblage of relative positions is called the Configuration of the system."

220. Kline, *Mathematics: The Loss of Certainty*, p. 136.

221. Boyer, *op. cit.*, p. 210.

222. Robinson, *op. cit.*, p. 71.

223. Kline, *Calculus II, op. cit.*, p. 366.

224. *Ibid.*, p. 367.

225. Henri Lebesgue, *op. cit.* p. 184.

226. Kline, Calculus II, *op. cit.*, p. 358.

227. Jesse Douglas, *Survey of The Theory Of Integration*, (New York: Scripta Mathematica, Yeshival College, 1941), p. 15.

228. *Ibid.*, p. 19-25.

229. The idea is to take the supposed denumerable infinity of points within an interval a-b and represent them by x_1, x_2, x_3, . . . x_n. Then cover each of these points by a sub-interval of length $\varepsilon/2$, where ε, of course, is a positive number which may be made as small as possible. x_1 is in the middle of a $\varepsilon/2$ sub-interval, such that it extends from $x_1 - \frac{1}{2}(\varepsilon/2)$ to $x_1 + \frac{1}{2}(\varepsilon/2)$; the same with x_2, which extends from $x_2 - \frac{1}{2}(\varepsilon/2^2)$ to $x_2 + \frac{1}{2}(\varepsilon/2^2)$. The same with any point x_n; it is to covered by a sub-interval of length $\varepsilon/2_n$ so that its extension is from $x_n - \frac{1}{2}(\varepsilon/2^n)$ to $x_2 + \frac{1}{2}(\varepsilon/2^n)$. Adding them together, the sum is to be $\varepsilon/2 + \varepsilon/2^2 + \varepsilon/2^3 + . . . \varepsilon/2^n$ . . . . The sum of the resulting geometric progression is S = ε. And since ε may be made as small as possible, it approaches 0. The set of rational points, therefore has a measure of zero. And since by another argument which is not challenged, b - a =1. Therefore, the conclusion is drawn that the rational numbers equal 0, and, therefore, that the complementary set, the irrational numbers, must equal 1, for 1-0 = 1. (Cf., Lieber, *Infinity*, pp. 229-300; Douglas, *op. cit.*, p *25-26).*

The fallacy is this argument is two fold: 1st, the existence of a denumerable infinity was refuted in Chapter V. 2nd, 0 is never reached by the term of ε, only approximated. Altogether, an uncountable number of them would reach quite a sum, regardless of how small they are. 3rd, there could still be a tiny breach of continuity within any fraction, $\frac{1}{2}(\varepsilon/2^n)$. 4th, nothing about the numeric size of the terminals of the irrational numbers in the "set" can be inferred from this. 5th, rational and irrational numbers do not exhaust the points.

230. Leibniz to Johann Bernoulli [1690], quoted in Douglas M. Jesseph, "Leibniz on the Foundations of the Calculus: The Question of the Reality of Infinitesimal Magnitudes," p. 13. (http: muse.jhu.edu/demo/posc/6.1jesseph.html).

231. *Loc. cit.*

232. *Ibid.*, p. 17.

233. Ferreirós, *op. cit.*, p. 47.

234. *Ibid.*, p. 65.

235. *Ibid.*, p. 58.

236. Ptolemy, "Trigonometry," in *Greek Mathematics With An English Translation by Ivor Thomas* (Cambridge, MA: Harvard University Press, 1993), pp. 411-13.
237. Euclid, I, *op. cit.*, p. 190.
238. J. J. Callahan, *op. cit.*, pp. 128-29.
239. Ibid., p. 129. "The points may be fixed on one line and any corresponding points may be taken on the other, and all the points on the other line may be made corresponding points, and vice versa."
240. Turner and Hazelett, *op. cit.*, pp. 271-305.
241. Adler, *op. cit.*, pp. 199, 205.
242. Baron Gottfried Leibniz, quoted in Kline, *Mathematics: The Loss of Certainty*, pp. 138-9.
243. Leibniz, quoted in Jesseph, *op. cit.*, p. 8.
244. Waismann, *op. cit.*, p. 192.
245. Kline, *Mathematics; The Loss of Certainty*, p. 182.
246. H.W.B. Joseph, *op. cit.*, p. 136.
247. Ferreriós, *op. cit.*, p. 252.
248. John Cook Wilson, *op. cit. II*, p. 643.
249. Kline, *Mathematics; The Loss Of Certainty*, p. 229.
250. Kurt Gödel, *On Formally Undecidable Propositions of Principia Mathematica And Related Systems*, tr. B. Meltzer with Introduction by R.B. Braithwaite, (New York: Basic Books, Inc., Publishers, 1962), pp. 55, 62. He states that it has to do only with the proofs for discursive propositions and these have to be the type of: "Proposition V: To every recursive relation R(x1 . . . xn) there corresponds an n-place relation-sign r (with the free variables u1, u2, . . . un) such that for every n-tuple of numbers (x1 . . . xn) the following hold:" R (x1 . . . xn) Bew [Sb, etc., etc.] and R (x1 . . . Xn Bew [Neg Sb (etc., etc.,) Needless to say, the reader will find nothing that even remotely resembles such displays in my book.
251. Ferreirós, *op. cit.*, p. 390.
252. John Von Neumann, quoted in *Ibid.*, p. 389.
253. *Ibid.*, p. 392.
254. In projective geometry, the phenomenon of perspective with its finite horizon lines and finite points of origin on those lines is converted into a situation where the point of origin is made into a supposed point of infinity. In order to do this on the two-dimensional coordinate system, in place of the standard x-y coordinates, they use three coordinates to a point. When one of these is made into a zero, this becomes the basis for division by zero, which is illegitimate, since the zero in question is an "ideal point" which has no place in space. Points, in turn, are given unrealistic definitions. Thus, a practical idea is given an absurd grounding.
255. The fractional form has greater precision than the digital.
256. Michael Starbird, *The Joy of Thinking: The Beauty and Power of Classical Mathematical Ideas*, Lecture One, (Chantilly, VA: The Teaching Company, 2003), p. 12.

257. Pesic, *op. cit.*, pp. 85-94.

258. Newton, *op. cit.*, p. 8.

259. *Http://ingeb.org/songs/forwanto.html* (06/14/2005.)

260. Peter Erickson, *op. cit.*, pp. 176-77.

261. *Ibid.*, p. 234-6.

262. *Ibid.*, p. 236.

263. *Ibid.* pp. 221-246.

264. The exception is when the terminal point of the duration is the present moment, as when one says, "The United States of America has existed as a nation since its constitution was ratified."

265. *Ibid.*, pp. 237. **"Stanford:** Why wasn't error presented as a third division of non-existence? What error purports to be does not exist.

"Philosophus: The reason for this is that error is metaphysically different from the past, which once was but is no longer and also of the future which either must or may be. What is in these categories is non-existent, but not in the same way as a wrong answer. This is the case whether the person who is in error did it through his own or was the victim of a liar. Consider the latter, more complicated case. What is true of it is also true of the simpler case. The action of the liar himself comes under Fact through Existence and Non-Existence; the same for the victim while he accepts the lie. This includes all the psychological paraphernalia which goes with it. As an act of imagination and will, it is a product of a mind. Its communication through some material means, such as the air or the written word, is included under Matter. It also takes place in Space and Time, etc.

"Stanford: If it is not the committing of a fallacy or the acceptance of a lie that makes it a non-fact, these being completely within the domain of fact, nor the mind's entertaining them as ideas or even as believing them (for states of consciousness too are facts), nor the content of the mistake or lie (these too existing in the mind), what then?

"Philosophus: You yourself almost said it, earlier. What is not a fact is what the falsehood purports to be. What is it that it purports to be? A *fact*. In short, error itself is a misapprehension. It comes under mind. The non-fact is what it would be if it were *apprehended*.

"Doxa: What if one were right for the wrong reasons?

"Philosophus: It would be a mistake all the same. Knowledge is not the agreement of one's notion of something with that something. It consists in the apprehension of it. The same for true opinion. It is not simply the content of that opinion that counts, but the grounds on which it is based.

"Stanford: Is error the only kind of non-fact?

"Philosophus: Two other sets of phenomena constitute non-fact rather than non-existence. They are what the content of fiction and of dreams are taken to be while they are being believed—not the images, but the conviction that they are real. The intention of the author of a fiction differs from that of a liar in that there is no attempt to sabotage anyone's consciousness. Furthermore, the consumer has voluntarily placed himself under a temporary suspension of judgment. Much of what is coming into his mind in such a situation fits under Existence and Non-Existence. That which does not fit under these are what they would be if they were real. What has just been said about deliberate fiction would also be the case with those moments of dreaming, whether in sleep, or in listening to music, or when one "loses one's self" in a day dream. Error, dreams, and fiction, are species of imagination. But the act of imagining, the content, the images, the feelings, the believing, etc.—all these are facts. What is not a fact is that any of the three are, or were, or will be apprehendable. They are not facts."

266. John Stuart Mill, *An Examination of Sir William Hamilton's Philosophy and of the Principal Philosophical Questions Discussed in his Writings*, ed. J. M. Robson, (Toronto, Canada: University Of Toronto Press, 1979), p. 4.

267. Robert Doughton, M.D., quoted in "Hypertension treatment proves 88 percent effective in Brazilian Study", *HSI Health Sciences Institute*, p. 6.

268. Turner and Hazelett, *op. cit.*, p. 3-17.

269. Ferrar Fenton, *The Holy Bible In Modern English, Containing The Complete Sacred Scriptures Of The Old And New Testaments Translated Into English Direct From The Original Hebrew, Chaldee And Greek*, (Merrimac, MA: Destiny Publishers, 1966, originally published in 1906), *passim*.

270. John Stuart Mill, editorial note no. 38 in James Mill's *Analysis Of The Phenomena Of The Human Mind, Ed. With Additional Notes By John Stuart Mill, Vol. I*, (London: Longmans, Green, Reader, and Dyer, 1878), pp. 413-423.

271. John Stuart Mill, editorial note no. 38, *Ibid., Vol. II*, pp. 198-9.

272. Some people object that it is wrong to call it an "idea," since no contemplation is needed in order to arrive at Time. But this is not the case with an idea which we did not make ourselves, but were born with. It is already there. This idea has always seemed a little mysterious, since people suppose it must have come from external experience, but cannot account for it there.

273. Wilson, *Ibid.*, p. 516.

274. Stephen Gaukroger, *Descartes An Intellectual Biography*, (Oxford, England: Clarendon Press, 1995), p. 350.

275. Adler, *op. cit.*, p. 266.

276. Wilson, *op. cit*, p. 78.

277. *Ibid.*, p. 74.

278. E.T. Bell, *op. cit.*, pp. 24-25.

279. *Ibid.*, p. 24.

280. Gerhard Kraus, *Has Hawking Erred?* (London, England: Janus Publishing Co.), 1993, pp. 55-56.

281. Albert Einstein, *Relativity: The Special and the General Theory*, tr. Robert Lawson, (New York: Crown Publishers, Inc.,1960), p. 155.

282. The premise behind the famous Dedekind cut was that the common irrationals are breaches in continuity.

283. Girolamo Cardano, *Ars Magna or The Rules of Algebra*, tr. by T. Richard Witmer, (New York: Dover Publications, Inc., 1993), pp.76-77.

284. Bernhard Riemann, quoted in Bell, *op. cit.* p. 491.

285. Ralph Waldo Emerson, quoted by Robert Welch in *The Romance Of Education*, (Belmont, MA: Western Islands, 1973), p. 209.

286. "So long as the direction in the four-dimensional world is space-like, no difficulty arises. But when we pass over to time-like directions (within the cone of absolute past or future) the directed radius is an imaginary length. Unless the object ignores the warning symbol of √-1 it has no standard of reference for settling its time extension." (Eddington, *op. cit.*, p. 146.)

287. Against this, it might be argued that negative square roots signify rotation; that rotation is a type of motion, and motion is as fundamental as space and cannot be reduced to it; and furthermore, negative square roots were found as solutions of algebraic equations before Descartes' invention of coordinate geometry. And therefore, since it only signifies rotation, the negative number inside the radical need not be veritable.

    The answer is first, directions are abstracted from lines; numbers are abstracted from the lengths of lines, and length is an attribute of space. Although rotation and space are not reducible in terms of each other, it is still a fact that although there can be space without rotation, there can be no rotation without space.

    Second, the veritable provides the extension for the imaginary line. Therefore, since the veritable is intrinsic to the imaginary and since spacial considerations predominate, it is the veritable that is symbolized; it is the mid-point in the radius vector's arc between +1 and -1 on the real scale.

    Third, √-1 does not mean rotation by itself. It is when the real and the veritable are simultaneously used in the same axis and then set against by a real axis that rotation is indicated.

288. Aristotle, *Physics*, Bk IV, 220b, 15, *op. cit*, p. 294.

289. Gaukroger, *op. cit.*, p. 367.

290. Albert Einstein, *Ideas and Opinions*, quoted in Henry H. Lindner's "Beyond Consciousness to Cosmos—Beyond Relativity and Quantum Theory to Cosmic Theory," *Physics Essays*, Vo. 15, No. 1, 2001, p. 122

291. Albert Einstein, *Relativity, The Spacial and General Theory*, (Crown Publishers, Inc.: New York, 1961), p. 10.

292. Albert Einstein, in *Albert Einstein: Philosopher-Scientist*, quoted in Lindner, *loc. cit.*

293. *Loc. cit.*

294. *Ibid.*, p. 114.

295. Albert Einstein, *Ideas and Opinions*, quoted in Ibid., p. 122.

296. Linder, *op. cit.*, p. 122.

297. *Ibid.*, pp. 113, 119.

298. Mott-Smith, *op. cit.*, p. 77.

299. Lindner, *op. cit.*, p. 118. "Einstein's Relativity and the QT were thus faithful implementations of Bishop Berkeley's vision of Science. They merely described the observer's experience with no reference to any physical Cosmos or causes—as if humans were experiencing a shared hallucination and wanted only to discover its rules."

300. J. Robert Oppenheimer, http://www.brainyquote.com/quotes/quotes/j/jrobertop.106068.html (06/14/05.)

301. Wu Shuping, *Four Books and Five Jing in Modern Chinese* (Beijing International Cultural Publication, Inc., 1996) Vol. II, *Yi Chiing*, p. 137; quoted by Samuel Wang and Ethel R. Nelson in *God and the Ancient Chinese*, (Dunlap, TN: Read Books Publisher, 1998), p. 98.

302. *Loc. cit.*

303. Wang and Nelson, *op. cit.*, pp. 218-219, 233, 265.

304. Ralph Baierlein, *Thermal Physics*, Cambridge, (UK: Cambridge University Press, 1999), pp. 337-38.

305. *Ibid.*, p. 352.

306. *Ibid.*, p. 333.

307. *Ibid.*, p. 349.

308. *Ibid.*, p. 345.

309. *Ibid.*, p. 345.

310. *Ibid.*, p. 347.

311. Cf., *Ibid.*, p. 177.

312. Mott-Smith, *op. cit.*,p. 91.

INDEX OF IMPORTANT
SUBJECTS AND PERSONS

Because of their high frequency of occurrence, not every use of certain important words is included. These are words such as: "finite," "infinitesimal," and "unit." Persons listed are those who figured in the argument; they are not always famous or great contributors to human thought. Names which appear only in the Endnotes or in the Bibliography are usually not listed here.

A

Asymptote,
—as Potential Infinity, 26-8, 82, 91, 220, 228-30.
—Complete-able, 81-2, 102, 187.
Axioms
—Definition, cf., Universal.
—Geometry, 60-3, 109, 113-4, 173-4, 177, 180.
Axis
—Imaginary, CHAPTERS XIII-XV, 221, 247, 254.
—Real, CHAPTERS XIII-XV, 220, 254.
—Veritable, CHAPTERS XIII-XV, 254.

B

Baierlein, Ralph, 228-9, 231, 255.
Barr, Stephen, 32, 231, 240.
Bayle, Pierre, 176.
Bell, Stoughton, cf., Tangent.
Berkeley, Bishop George, 19, 30, 226, 234.
Bernoulli, Johann, 168, 250.
Bolzano, Bernard, 36-7, 47, 63.
Boole, George, cf., Numbers—Imaginary.
Borchardt, Glenn, 89, 231, 243-4.
Boyer, Carl, 38, 89, 104, 147, 231, 241-2, 248-50, 258.
Buddhism, 226.
Butler, Rhett, cf., Non-Euclidean.

C

Caesar, Julius, cf., Infinitesimal—of Time.
Callahan, Jeremiah J., 90, 171-2, 174, 232, 244, 251.
Cantor, Georg, 10-11, 13, 35, 38, 44, 49, 51, 54, 56, 80-3, 86, 215-6, 232, 242,
   244.
Cardano, Girolamo, 214, 232, 254.
Cartesian Coordinates, CHAPTERS XIII-XV, 246-7.
Cauchy, Augustin Louis, 30, 147, 248-9.
Circle, 16-7, 25, 30-1, 59, 61, 63, 67, 82, 94-7, 105-7, 112, 116-7, 122-3, 126,
   128, 130, 132, 142-3, 171, 173-5, 178-9, 190-1, 193, 217-8, 220, 234, 243,
   247.
Combination Lock, cf., Numbers—Veritable.
Confucius, 226.

E

F

G

J

K

L

M

P

Q

R

U

V

W

Y

Z